IECC®

INTERNATIONAL ENERGY CONSERVATION CODE®

2009

Receive **FREE** updates, excerpts of code references, technical articles, and more when you register your code book. Go to

www.iccsafe.org/CodesPlus today!

2009 International Energy Conservation Code®

First Printing: January 2009
Second Printing: March 2009
Third Printing: August 2009
Fourth Printing: February 2010
Fifth Printing: May 2010
Sixth Printing: July 2010
Seventh Printing: October 2011
Eighth Printing: February 2012
Ninth Printing: April 2012
Tenth Printing: February 2013

ISBN: 978-1-58001-742-8 (soft-cover edition)

38336-T015897

PRINTED IN THE U.S.A.

PREFACE

Introduction

Internationally, code officials recognize the need for a modern, up-to-date energy conservation code addressing the design of energy-efficient building envelopes and installation of energy efficient mechanical, lighting and power systems through requirements emphasizing performance. The *International Energy Conservation Code*®, in this 2009 edition, is designed to meet these needs through model code regulations that will result in the optimal utilization of fossil fuel and nondepletable resources in all communities, large and small.

This comprehensive energy conservation code establishes minimum regulations for energy efficient buildings using prescriptive and performance-related provisions. It is founded on broad-based principles that make possible the use of new materials and new energy efficient designs. This 2009 edition is fully compatible with all the *International Codes*® (I-Codes®) published by the International Code Council (ICC)®, including: the *International Building Code*®, *International Existing Building Code*®, *International Fire Code*®, *International Fuel Gas Code*®, *International Mechanical Code*®, ICC *Performance Code*®, *International Plumbing Code*®, *International Private Sewage Disposal Code*®, *International Property Maintenance Code*®, *International Residential Code*®, *International Wildland-Urban Interface Code*™ and *International Zoning Code*®.

The *International Energy Conservation Code* provisions provide many benefits, among which is the model code development process that offers an international forum for energy professionals to discuss performance and prescriptive code requirements. This forum provides an excellent arena to debate proposed revisions. This model code also encourages international consistency in the application of provisions.

Development

The first edition of the *International Energy Conservation Code* (1998) was based on the 1995 edition of the *Model Energy Code* promulgated by the Council of American Building Officials (CABO) and included changes approved through the CABO Code Development Procedures through 1997. CABO assigned all rights and responsibilities to the International Code Council and its three statutory members at that time, including Building Officials and Code Administrators International, Inc. (BOCA), International Conference of Building Officials (ICBO) and Southern Building Code Congress International (SBCCI). This 2009 edition presents the code as originally issued, with changes reflected in the 2000, 2003 and 2006 editions and further changes approved through the ICC Code Development Process through 2008. A new edition such as this is promulgated every three years.

This code is founded on principles intended to establish provisions consistent with the scope of an energy conservation code that adequately conserves energy; provisions that do not unnecessarily increase construction costs; provisions that do not restrict the use of new materials, products or methods of construction; and provisions that do not give preferential treatment to particular types or classes of materials, products or methods of construction.

Adoption

The *International Energy Conservation Code* is available for adoption and use by jurisdictions internationally. Its use within a governmental jurisdiction is intended to be accomplished through adoption by reference in accordance with proceedings establishing the jurisdiction's laws. At the time of adoption, jurisdictions should insert the appropriate information in provisions requiring specific local information, such as the name of the adopting jurisdiction. These locations are shown in bracketed words in small capital letters in the code and in the sample ordinance. The sample adoption ordinance on page vii addresses several key elements of a code adoption ordinance, including the information required for insertion into the code text.

Maintenance

The *International Energy Conservation Code* is kept up to date through the review of proposed changes submitted by code enforcing officials, industry representatives, design professionals and other interested parties. Proposed changes are carefully considered through an open code development process in which all interested and affected parties may participate.

The contents of this work are subject to change both through the Code Development Cycles and the governmental body that enacts the code into law. For more information regarding the code development process, contact the Code and Standard Development Department of the International Code Council.

While the development procedure of the *International Energy Conservation Code* assures the highest degree of care, ICC, its members and those participating in the development of this code do not accept any liability resulting from compliance or noncompliance with the provisions because ICC and its members do not have the power or authority to police or enforce compliance with the contents of this code. Only the governmental body that enacts the code into law has such authority.

Marginal Markings

Solid vertical lines in the margins within the body of the code indicate a technical change from the requirements of the 2006 edition. Deletion indicators in the form of an arrow (➡) are provided in the margin where an entire section, paragraph, exception or table has been deleted or an item in a list of items or a table has been deleted.

Italicized Terms

Selected terms set forth in Chapter 2, Definitions, are italicized where they appear in code text. Such terms are not italicized where the definition set forth in Chapter 2 does not impart the intended meaning in the use of the term. The terms selected have definitions which the user should read carefully to facilitate better understanding of the code.

Effective Use of the International Energy Conservation Code

The *International Energy Conservation Code* (IECC) is a model code that regulates minimum energy conservation requirements for new buildings. The IECC addresses energy conservation requirements for all aspects of energy uses in both commercial and residential construction, including heating and ventilating, lighting, water heating, and power usage for appliances and building systems.

The IECC is a design document. For example, before one constructs a building, the designer must determine the minimum insulation *R*-values and fenestration *U*-factors for the building exterior envelope. Depending on whether the building is for residential use or for commercial use, the IECC sets forth minimum requirements for exterior envelope insulation, window and door *U*-factors and SHGC ratings, duct insulation, lighting and power efficiency, and water distribution insulation.

Arrangement and Format of the 2009 IECC

Before applying the requirements of the IECC it is beneficial to understand its arrangement and format. The IECC, like other codes published by ICC, is arranged and organized to follow sequential steps that generally occur during a plan review or inspection. The IECC is divided into five different parts:

Chapters	Subjects
1–2	Administration and definitions
3	Climate zones and general materials requirements
4	Energy efficiency for residential buildings
5	Energy efficiency for commercial buildings
6	Referenced standards

The following is a chapter-by-chapter synopsis of the scope and intent of the provisions of the *International Energy Conservation Code*:

Chapter 1 Administration. This chapter contains provisions for the application, enforcement and administration of subsequent requirements of the code. In addition to establishing the scope of the code, Chapter 1 identifies which buildings and structures come under its purview. Chapter 1 is largely concerned with maintaining "due process of law" in enforcing the energy conservation criteria contained in the body of the code. Only through careful observation of the administrative provisions can the building official reasonably expect to demonstrate that "equal protection under the law" has been provided.

Chapter 2 Definitions. All terms that are defined in the code are listed alphabetically in Chapter 2. While a defined term may be used in one chapter or another, the meaning provided in Chapter 2 is applicable throughout the code.

Additional definitions regarding climate zones are found in Tables 301.3(1) and (2). These are not listed in Chapter 2.

Where understanding of a term's definition is especially key to or necessary for understanding of a particular code provision, the term is show in *italics* wherever it appears in the code. This is true only for those terms that have a meaning that is unique to the code. In other words, the generally understood meaning of a term or phrase might not be sufficient or consistent with the meaning prescribed by the code; therefore, it is essential that the code-defined meaning be known.

Guidance regarding tense, gender and plurality of defined terms as well as guidance regarding terms not defined in this code is provided.

Chapter 3 Climate Zones. Chapter 3 specifies the climate zones that will serve to establish the exterior design conditions. In addition, Chapter 3 provides interior design conditions that are used as a basis for assumptions in heating and cooling load calculations, and provides basic material requirements for insulation materials and fenestration materials.

Climate has a major impact on the energy use of most buildings. The code establishes many requirements such as wall and roof insulation *R*-values, window and door thermal transmittance requirement (*U*-factors) as well as provisions that affect the mechanical systems based upon the climate where the building is located. This chapter will contain the information that will be used to properly assign the building location into the correct climate zone and will then be used as the basis for establishing requirements or elimination of requirements.

Chapter 4 Residential Energy Efficiency. Chapter 4 contains the energy-efficiency-related requirements for the design and construction of residential buildings regulated under this code. It should be noted that the definition of a *residential building* in this code is unique for this code. In this code, a *residential building* is an R-2, R-3 or R-4 building three stories or less in height. All other R-1 buildings, including residential buildings greater than three stories in height, are regulated by the energy conservation requirements of Chapter 5. The applicable portions of a residential building must comply with the provisions within this chapter for energy efficiency. This chapter defines requirements for the portions of the building and building systems that impact energy use in new residential construction and promotes the effective use of energy. The provisions within the chapter promote energy efficiency in the building envelope, the heating and cooling system and the service water heating system of the building.

Chapter 5 Commercial Energy Efficiency. Chapter 5 contains the energy-efficiency-related requirements for the design and construction of most types of commercial buildings and residential buildings greater than three stories in height above grade. Residential buildings, townhouses and garden apartments three stories or less in height are covered in Chapter 4. Like Chapter 4, this chapter defines requirements for the portions of the building and building systems that impact energy use in new commercial construction and new residential construction greater than three stories in height, and promotes the effective use of energy. The provisions within the chapter promote energy efficiency in the building envelope, the heating and cooling system and the service water heating system of the building.

Chapter 6 Referenced Standards. The code contains numerous references to standards that are used to regulate materials and methods of construction. Chapter 6 contains a comprehensive list of all standards that are referenced in the code. The standards are part of the code to the extent of the reference to the standard. Compliance with the referenced standard is necessary for compliance with this code. By providing specifically adopted standards, the construction and installation requirements necessary for compliance with the code can be readily determined. The basis for code compliance is, therefore, established and available on an equal basis to the code official, contractor, designer and owner.

Chapter 6 is organized in a manner that makes it easy to locate specific standards. It lists all of the referenced standards, alphabetically, by acronym of the promulgating agency of the standard. Each agency's standards are then listed in either alphabetical or numeric order based upon the standard identification. The list also contains the title of the standard; the edition (date) of the standard referenced; any addenda included as part of the ICC adoption; and the section or sections of this code that reference the standard.

ORDINANCE

The International Codes are designed and promulgated to be adopted by reference by ordinance. Jurisdictions wishing to adopt the 2009 *International Energy Conservation Code* as an enforceable regulation governing energy efficient building envelopes and installation of energy efficient mechanical, lighting and power systems should ensure that certain factual information is included in the adopting ordinance at the time adoption is being considered by the appropriate governmental body. The following sample adoption ordinance addresses several key elements of a code adoption ordinance, including the information required for insertion into the code text.

SAMPLE ORDINANCE FOR ADOPTION OF THE *INTERNATIONAL ENERGY CONSERVATION CODE* ORDINANCE NO._____

An ordinance of the **[JURISDICTION]** adopting the 2009 edition of the *International Energy Conservation Code*, regulating and governing energy efficient building envelopes and installation of energy efficient mechanical, lighting and power systems in the **[JURISDICTION]**; providing for the issuance of permits and collection of fees therefor; repealing Ordinance No. _____ of the **[JURISDICTION]** and all other ordinances and parts of the ordinances in conflict therewith.

The **[GOVERNING BODY]** of the **[JURISDICTION]** does ordain as follows:

Section 1. That a certain document, three (3) copies of which are on file in the office of the **[TITLE OF JURISDICTION'S KEEPER OF RECORDS]** of **[NAME OF JURISDICTION]**, being marked and designated as the *International Energy Conservation Code*, 2009 edition, as published by the International Code Council, be and is hereby adopted as the Energy Conservation Code of the **[JURISDICTION]**, in the State of **[STATE NAME]** for regulating and governing energy efficient building envelopes and installation of energy efficient mechanical, lighting and power systems as herein provided; providing for the issuance of permits and collection of fees therefor; and each and all of the regulations, provisions, penalties, conditions and terms of said Energy Conservation Code on file in the office of the **[JURISDICTION]** are hereby referred to, adopted, and made a part hereof, as if fully set out in this ordinance, with the additions, insertions, deletions and changes, if any, prescribed in Section 2 of this ordinance.

Section 2. The following sections are hereby revised:

Section 101.1. Insert: **[NAME OF JURISDICTION]**.

Section 108.4. Insert: **[DOLLAR AMOUNT]** in two places.

Section 3. That Ordinance No. _____ of **[JURISDICTION]** entitled **[FILL IN HERE THE COMPLETE TITLE OF THE ORDINANCE OR ORDINANCES IN EFFECT AT THE PRESENT TIME SO THAT THEY WILL BE REPEALED BY DEFINITE MENTION]** and all other ordinances or parts of ordinances in conflict herewith are hereby repealed.

Section 4. That if any section, subsection, sentence, clause or phrase of this ordinance is, for any reason, held to be unconstitutional, such decision shall not affect the validity of the remaining portions of this ordinance. The **[GOVERNING BODY]** hereby declares that it would have passed this ordinance, and each section, subsection, clause or phrase thereof, irrespective of the fact that any one or more sections, subsections, sentences, clauses and phrases be declared unconstitutional.

Section 5. That nothing in this ordinance or in the *International Energy Conservation Code*® hereby adopted shall be construed to affect any suit or proceeding impending in any court, or any rights acquired, or liability incurred, or any cause or causes of action acquired or existing, under any act or ordinance hereby repealed as cited in Section 3 of this ordinance; nor shall any just or legal right or remedy of any character be lost, impaired or affected by this ordinance.

Section 6. That the **[JURISDICTION'S KEEPER OF RECORDS]** is hereby ordered and directed to cause this ordinance to be published. (An additional provision may be required to direct the number of times the ordinance is to be published and to specify that it is to be in a newspaper in general circulation. Posting may also be required.)

Section 7. That this ordinance and the rules, regulations, provisions, requirements, orders and matters established and adopted hereby shall take effect and be in full force and effect **[TIME PERIOD]** from and after the date of its final passage and adoption.

TABLE OF CONTENTS

CHAPTER 1

ADMINISTRATION

■ PART 1—SCOPE AND APPLICATION

SECTION 101
SCOPE AND GENERAL REQUIREMENTS

101.1 Title. This code shall be known as the *International Energy Conservation Code* of [NAME OF JURISDICTION], and shall be cited as such. It is referred to herein as "this code."

101.2 Scope. This code applies to *residential* and *commercial buildings*.

101.3 Intent. This code shall regulate the design and construction of buildings for the effective use of energy. This code is intended to provide flexibility to permit the use of innovative approaches and techniques to achieve the effective use of energy. This code is not intended to abridge safety, health or environmental requirements contained in other applicable codes or ordinances.

101.4 Applicability. Where, in any specific case, different sections of this code specify different materials, methods of construction or other requirements, the most restrictive shall govern. Where there is a conflict between a general requirement and a specific requirement, the specific requirement shall govern.

101.4.1 Existing buildings. Except as specified in this chapter, this code shall not be used to require the removal, *alteration* or abandonment of, nor prevent the continued use and maintenance of, an existing building or building system lawfully in existence at the time of adoption of this code.

101.4.2 Historic buildings. Any building or structure that is listed in the State or National Register of Historic Places; designated as a historic property under local or state designation law or survey; certified as a contributing resource with a National Register listed or locally designated historic district; or with an opinion or certification that the property is eligible to be listed on the National or State Registers of Historic Places either individually or as a contributing building to a historic district by the State Historic Preservation Officer or the Keeper of the National Register of Historic Places, are exempt from this code.

101.4.3 Additions, alterations, renovations or repairs. Additions, alterations, renovations or repairs to an existing building, building system or portion thereof shall conform to the provisions of this code as they relate to new construction without requiring the unaltered portion(s) of the existing building or building system to comply with this code. Additions, alterations, renovations or repairs shall not create an unsafe or hazardous condition or overload existing building systems. An addition shall be deemed to comply with this code if the addition alone complies or if the exist-ing building and addition comply with this code as a single building.

> **Exception:** The following need not comply provided the energy use of the building is not increased:
>
> 1. Storm windows installed over existing fenestration.
>
> 2. Glass only replacements in an existing sash and frame.
>
> 3. Existing ceiling, wall or floor cavities exposed during construction provided that these cavities are filled with insulation.
>
> 4. Construction where the existing roof, wall or floor cavity is not exposed.
>
> 5. Reroofing for roofs where neither the sheathing nor the insulation is exposed. Roofs without insulation in the cavity and where the sheathing or insulation is exposed during reroofing shall be insulated either above or below the sheathing.
>
> 6. Replacement of existing doors that separate *conditioned space* from the exterior shall not require the installation of a vestibule or revolving door, provided, however, that an existing vestibule that separates a *conditioned space* from the exterior shall not be removed,
>
> 7. Alterations that replace less than 50 percent of the luminaires in a space, provided that such alterations do not increase the installed interior lighting power.
>
> 8. Alterations that replace only the bulb and ballast within the existing luminaires in a space provided that the *alteration* does not increase the installed interior lighting power.

101.4.4 Change in occupancy or use. Spaces undergoing a change in occupancy that would result in an increase in demand for either fossil fuel or electrical energy shall comply with this code. Where the use in a space changes from one use in Table 505.5.2 to another use in Table 505.5.2, the installed lighting wattage shall comply with Section 505.5.

101.4.5 Change in space conditioning. Any nonconditioned space that is altered to become *conditioned space* shall be required to be brought into full compliance with this code.

101.4.6 Mixed occupancy. Where a building includes both *residential* and *commercial* occupancies, each occupancy shall be separately considered and meet the applicable provisions of Chapter 4 for *residential* and Chapter 5 for *commercial*.

101.5 Compliance. *Residential buildings* shall meet the provisions of Chapter 4. *Commercial buildings* shall meet the provisions of Chapter 5.

101.5.1 Compliance materials. The *code official* shall be permitted to approve specific computer software, worksheets, compliance manuals and other similar materials that meet the intent of this code.

101.5.2 Low energy buildings. The following buildings, or portions thereof, separated from the remainder of the building by *building thermal envelope* assemblies complying with this code shall be exempt from the *building thermal envelope* provisions of this code:

1. Those with a peak design rate of energy usage less than 3.4 Btu/h·ft^2 (10.7 W/m^2) or 1.0 watt/ft^2 (10.7 W/m^2) of floor area for space conditioning purposes.

2. Those that do not contain *conditioned space*.

SECTION 102
ALTERNATE MATERIALS—METHOD OF CONSTRUCTION, DESIGN OR INSULATING SYSTEMS

102.1 General. This code is not intended to prevent the use of any material, method of construction, design or insulating system not specifically prescribed herein, provided that such construction, design or insulating system has been *approved* by the *code official* as meeting the intent of this code.

102.1.1 Above code programs. The *code official* or other authority having jurisdiction shall be permitted to deem a national, state or local energy efficiency program to exceed the energy efficiency required by this code. Buildings *approved* in writing by such an energy efficiency program shall be considered in compliance with this code. The requirements identified as "mandatory" in Chapters 4 and 5 of this code, as applicable, shall be met.

PART 2—ADMINISTRATION AND ENFORCEMENT

SECTION 103
CONSTRUCTION DOCUMENTS

103.1 General. Construction documents and other supporting data shall be submitted in one or more sets with each application for a permit. The construction documents shall be prepared by a registered design professional where required by the statutes of the jurisdiction in which the project is to be constructed. Where special conditions exist, the *code official* is authorized to require necessary construction documents to be prepared by a registered design professional.

Exception: The *code official* is authorized to waive the requirements for construction documents or other supporting data if the *code official* determines they are not necessary to confirm compliance with this code.

103.2 Information on construction documents. Construction documents shall be drawn to scale upon suitable material. Electronic media documents are permitted to be submitted when *approved* by the *code official*. Construction documents shall be of sufficient clarity to indicate the location, nature and extent of the work proposed, and show in sufficient detail pertinent data and features of the building, systems and equipment as herein governed. Details shall include, but are not limited to, as applicable, insulation materials and their *R*-values; fenestration *U*-factors and SHGCs; area-weighted *U*-factor and SHGC calculations; mechanical system design criteria; mechanical and service water heating system and equipment types, sizes and efficiencies; economizer description; equipment and systems controls; fan motor horsepower (hp) and controls; duct sealing, duct and pipe insulation and location; lighting fixture schedule with wattage and control narrative; and air sealing details.

103.3 Examination of documents. The *code official* shall examine or cause to be examined the accompanying construction documents and shall ascertain whether the construction indicated and described is in accordance with the requirements of this code and other pertinent laws or ordinances.

103.3.1 Approval of construction documents. When the *code official* issues a permit where construction documents are required, the construction documents shall be endorsed in writing and stamped "Reviewed for Code Compliance." Such *approved* construction documents shall not be changed, modified or altered without authorization from the *code official*. Work shall be done in accordance with the *approved* construction documents.

One set of construction documents so reviewed shall be retained by the *code official*. The other set shall be returned to the applicant, kept at the site of work and shall be open to inspection by the *code official* or a duly authorized representative.

103.3.2 Previous approvals. This code shall not require changes in the construction documents, construction or designated occupancy of a structure for which a lawful permit has been heretofore issued or otherwise lawfully authorized, and the construction of which has been pursued in good faith within 180 days after the effective date of this code and has not been abandoned.

103.3.3 Phased approval. The *code official* shall have the authority to issue a permit for the construction of part of an energy conservation system before the construction documents for the entire system have been submitted or *approved*, provided adequate information and detailed statements have been filed complying with all pertinent requirements of this code. The holders of such permit shall proceed at their own risk without assurance that the permit for the entire energy conservation system will be granted.

103.4 Amended construction documents. Changes made during construction that are not in compliance with the *approved* construction documents shall be resubmitted for approval as an amended set of construction documents.

103.5 Retention of construction documents. One set of *approved* construction documents shall be retained by the *code official* for a period of not less than 180 days from date of completion of the permitted work, or as required by state or local laws.

SECTION 104
INSPECTIONS

104.1 General. Construction or work for which a permit is required shall be subject to inspection by the *code official*.

104.2 Required approvals. Work shall not be done beyond the point indicated in each successive inspection without first obtaining the approval of the *code official*. The *code official*, upon notification, shall make the requested inspections and shall either indicate the portion of the construction that is satisfactory as completed, or notify the permit holder or his or her agent wherein the same fails to comply with this code. Any portions that do not comply shall be corrected and such portion shall not be covered or concealed until authorized by the *code official*.

104.3 Final inspection. The building shall have a final inspection and not be occupied until *approved*.

104.4 Reinspection. A building shall be reinspected when determined necessary by the *code official*.

104.5 Approved inspection agencies. The *code official* is authorized to accept reports of *approved* inspection agencies, provided such agencies satisfy the requirements as to qualifications and reliability.

104.6 Inspection requests. It shall be the duty of the holder of the permit or their duly authorized agent to notify the *code official* when work is ready for inspection. It shall be the duty of the permit holder to provide access to and means for inspections of such work that are required by this code.

104.7 Reinspection and testing. Where any work or installation does not pass an initial test or inspection, the necessary corrections shall be made so as to achieve compliance with this code. The work or installation shall then be resubmitted to the *code official* for inspection and testing.

104.8 Approval. After the prescribed tests and inspections indicate that the work complies in all respects with this code, a notice of approval shall be issued by the *code official*.

> **104.8.1 Revocation.** The *code official* is authorized to, in writing, suspend or revoke a notice of approval issued under the provisions of this code wherever the certificate is issued in error, or on the basis of incorrect information supplied, or where it is determined that the building or structure, premise, or portion thereof is in violation of any ordinance or regulation or any of the provisions of this code.

SECTION 105
VALIDITY

105.1 General. If a portion of this code is held to be illegal or void, such a decision shall not affect the validity of the remainder of this code.

SECTION 106
REFERENCED STANDARDS

106.1 General. The codes and standards referenced in this code shall be those listed in Chapter 6, and such codes and standards shall be considered as part of the requirements of this code to the prescribed extent of each such reference.

106.2 Conflicting requirements. Where the provisions of this code and the referenced standards conflict, the provisions of this code shall take precedence.

106.3 Application of references. References to chapter or section numbers, or to provisions not specifically identified by number, shall be construed to refer to such chapter, section or provision of this code.

106.4 Other laws. The provisions of this code shall not be deemed to nullify any provisions of local, state or federal law.

SECTION 107
FEES

107.1 Fees. A permit shall not be issued until the fees prescribed in Section 107.2 have been paid, nor shall an amendment to a permit be released until the additional fee, if any, has been paid.

107.2 Schedule of permit fees. A fee for each permit shall be paid as required, in accordance with the schedule as established by the applicable governing authority.

107.3 Work commencing before permit issuance. Any person who commences any work before obtaining the necessary permits shall be subject to an additional fee established by the *code official*, which shall be in addition to the required permit fees.

107.4 Related fees. The payment of the fee for the construction, *alteration*, removal or demolition of work done in connection to or concurrently with the work or activity authorized by a permit shall not relieve the applicant or holder of the permit from the payment of other fees that are prescribed by law.

107.5 Refunds. The *code official* is authorized to establish a refund policy.

SECTION 108
STOP WORK ORDER

108.1 Authority. Whenever the *code official* finds any work regulated by this code being performed in a manner either contrary to the provisions of this code or dangerous or unsafe, the *code official* is authorized to issue a stop work order.

108.2 Issuance. The stop work order shall be in writing and shall be given to the owner of the property involved, or to the owner's agent, or to the person doing the work. Upon issuance of a stop work order, the cited work shall immediately cease. The stop work order shall state the reason for the order, and the conditions under which the cited work will be permitted to resume.

108.3 Emergencies. Where an emergency exists, the *code official* shall not be required to give a written notice prior to stopping the work.

108.4 Failure to comply. Any person who shall continue any work after having been served with a stop work order, except such work as that person is directed to perform to remove a violation or unsafe condition, shall be liable to a fine of not less than [AMOUNT] dollars or more than [AMOUNT] dollars.

SECTION 109
BOARD OF APPEALS

109.1 General. In order to hear and decide appeals of orders, decisions or determinations made by the *code official* relative to the application and interpretation of this code, there shall be and is hereby created a board of appeals. The *code official* shall be an ex officio member of said board but shall have no vote on any matter before the board. The board of appeals shall be appointed by the governing body and shall hold office at its pleasure. The board shall adopt rules of procedure for conducting its business, and shall render all decisions and findings in writing to the appellant with a duplicate copy to the *code official*.

109.2 Limitations on authority. An application for appeal shall be based on a claim that the true intent of this code or the rules legally adopted thereunder have been incorrectly interpreted, the provisions of this code do not fully apply or an equally good or better form of construction is proposed. The board shall have no authority to waive requirements of this code.

109.3 Qualifications. The board of appeals shall consist of members who are qualified by experience and training and are not employees of the jurisdiction.

CHAPTER 2
DEFINITIONS

SECTION 201
GENERAL

201.1 Scope. Unless stated otherwise, the following words and terms in this code shall have the meanings indicated in this chapter.

201.2 Interchangeability. Words used in the present tense include the future; words in the masculine gender include the feminine and neuter; the singular number includes the plural and the plural includes the singular.

201.3 Terms defined in other codes. Terms that are not defined in this code but are defined in the *International Building Code, International Fire Code, International Fuel Gas Code, International Mechanical Code, International Plumbing Code* or the *International Residential Code* shall have the meanings ascribed to them in those codes.

201.4 Terms not defined. Terms not defined by this chapter shall have ordinarily accepted meanings such as the context implies.

SECTION 202
GENERAL DEFINITIONS

ABOVE-GRADE WALL. A wall more than 50 percent above grade and enclosing *conditioned space*. This includes between-floor spandrels, peripheral edges of floors, roof and basement knee walls, dormer walls, gable end walls, walls enclosing a mansard roof and skylight shafts.

ACCESSIBLE. Admitting close approach as a result of not being guarded by locked doors, elevation or other effective means (see "Readily *accessible*").

ADDITION. An extension or increase in the *conditioned space* floor area or height of a building or structure.

AIR BARRIER. Material(s) assembled and joined together to provide a barrier to air leakage through the building envelope. An air barrier may be a single material or a combination of materials.

ALTERATION. Any construction or renovation to an existing structure other than repair or addition that requires a permit. Also, a change in a mechanical system that involves an extension, addition or change to the arrangement, type or purpose of the original installation that requires a permit.

APPROVED. Approval by the *code official* as a result of investigation and tests conducted by him or her, or by reason of accepted principles or tests by nationally recognized organizations.

AUTOMATIC. Self-acting, operating by its own mechanism when actuated by some impersonal influence, as, for example, a change in current strength, pressure, temperature or mechanical configuration (see "Manual").

BASEMENT WALL. A wall 50 percent or more below grade and enclosing *conditioned space*.

BUILDING. Any structure used or intended for supporting or sheltering any use or occupancy.

BUILDING THERMAL ENVELOPE. The basement walls, exterior walls, floor, roof, and any other building element that enclose *conditioned space*. This boundary also includes the boundary between *conditioned space* and any exempt or unconditioned space.

C-FACTOR (THERMAL CONDUCTANCE). The coefficient of heat transmission (surface to surface) through a building component or assembly, equal to the time rate of heat flow per unit area and the unit temperature difference between the warm side and cold side surfaces (Btu/h ft^2 × °F) [W/(m^2 × K)].

CODE OFFICIAL. The officer or other designated authority charged with the administration and enforcement of this code, or a duly authorized representative.

COMMERCIAL BUILDING. For this code, all buildings that are not included in the definition of "Residential buildings."

CONDITIONED FLOOR AREA. The horizontal projection of the floors associated with the *conditioned space*.

CONDITIONED SPACE. An area or room within a building being heated or cooled, containing uninsulated ducts, or with a fixed opening directly into an adjacent *conditioned space*.

CRAWL SPACE WALL. The opaque portion of a wall that encloses a crawl space and is partially or totally below grade.

CURTAIN WALL. Fenestration products used to create an external nonload-bearing wall that is designed to separate the exterior and interior environments.

DAYLIGHT ZONE.

1. **Under skylights.** The area under skylights whose horizontal dimension, in each direction, is equal to the skylight dimension in that direction plus either the floor-to-ceiling height or the dimension to a ceiling height opaque partition, or one-half the distance to adjacent skylights or vertical fenestration, whichever is least.

2. **Adjacent to vertical fenestration.** The area adjacent to vertical fenestration which receives daylight through the fenestration. For purposes of this definition and unless more detailed analysis is provided, the daylight *zone* depth is assumed to extend into the space a distance of 15 feet (4572 mm) or to the nearest ceiling height opaque partition, whichever is less. The daylight *zone* width is assumed to be the width of the window plus 2 feet (610 mm) on each side, or the window width plus the distance to an opaque partition, or the window width plus one-half the distance to adjacent skylight or vertical fenestration, whichever is least.

DEMAND CONTROL VENTILATION (DCV). A ventilation system capability that provides for the automatic reduction of outdoor air intake below design rates when the actual occupancy of spaces served by the system is less than design occupancy.

DUCT. A tube or conduit utilized for conveying air. The air passages of self-contained systems are not to be construed as air ducts.

DUCT SYSTEM. A continuous passageway for the transmission of air that, in addition to ducts, includes duct fittings, dampers, plenums, fans and accessory air-handling equipment and appliances.

DWELLING UNIT. A single unit providing complete independent living facilities for one or more persons, including permanent provisions for living, sleeping, eating, cooking and sanitation.

ECONOMIZER, AIR. A duct and damper arrangement and automatic control system that allows a cooling system to supply outside air to reduce or eliminate the need for mechanical cooling during mild or cold weather.

ECONOMIZER, WATER. A system where the supply air of a cooling system is cooled indirectly with water that is itself cooled by heat or mass transfer to the environment without the use of mechanical cooling.

ENERGY ANALYSIS. A method for estimating the annual energy use of the *proposed design* and *standard reference design* based on estimates of energy use.

ENERGY COST. The total estimated annual cost for purchased energy for the building functions regulated by this code, including applicable demand charges.

ENERGY RECOVERY VENTILATION SYSTEM. Systems that employ air-to-air heat exchangers to recover energy from exhaust air for the purpose of preheating, precooling, humidifying or dehumidifying outdoor ventilation air prior to supplying the air to a space, either directly or as part of an HVAC system.

ENERGY SIMULATION TOOL. An *approved* software program or calculation-based methodology that projects the annual energy use of a building.

ENTRANCE DOOR. Fenestration products used for ingress, egress and access in nonresidential buildings, including, but not limited to, exterior entrances that utilize latching hardware and automatic closers and contain over 50-percent glass specifically designed to withstand heavy use and possibly abuse.

EXTERIOR WALL. Walls including both above-grade walls and basement walls.

FAN BRAKE HORSEPOWER (BHP). The horsepower delivered to the fan's shaft. Brake horsepower does not include the mechanical drive losses (belts, gears, etc.).

FAN SYSTEM BHP. The sum of the fan brake horsepower of all fans that are required to operate at fan system design conditions to supply air from the heating or cooling source to the *conditioned space(s)* and return it to the source or exhaust it to the outdoors.

FAN SYSTEM DESIGN CONDITIONS. Operating conditions that can be expected to occur during normal system operation that result in the highest supply fan airflow rate to conditioned spaces served by the system.

FAN SYSTEM MOTOR NAMEPLATE HP. The sum of the motor nameplate horsepower of all fans that are required to operate at design conditions to supply air from the heating or cooling source to the *conditioned space(s)* and return it to the source or exhaust it to the outdoors.

FENESTRATION. Skylights, roof windows, vertical windows (fixed or moveable), opaque doors, glazed doors, glazed block and combination opaque/glazed doors. Fenestration includes products with glass and nonglass glazing materials.

***F*-FACTOR.** The perimeter heat loss factor for slab-on-grade floors (Btu/h × ft × °F) [W/(m × K)].

HEAT TRAP. An arrangement of piping and fittings, such as elbows, or a commercially available heat trap that prevents thermosyphoning of hot water during standby periods.

HEATED SLAB. Slab-on-grade construction in which the heating elements, hydronic tubing, or hot air distribution system is in contact with, or placed within or under, the slab.

HIGH-EFFICACY LAMPS. Compact fluorescent lamps, T-8 or smaller diameter linear fluorescent lamps, or lamps with a minimum efficacy of:

1. 60 lumens per watt for lamps over 40 watts,

2. 50 lumens per watt for lamps over 15 watts to 40 watts, and

3. 40 lumens per watt for lamps 15 watts or less.

HUMIDISTAT. A regulatory device, actuated by changes in humidity, used for automatic control of relative humidity.

INFILTRATION. The uncontrolled inward air leakage into a building caused by the pressure effects of wind or the effect of differences in the indoor and outdoor air density or both.

INSULATING SHEATHING. An insulating board with a core material having a minimum *R*-value of R-2.

LABELED. Equipment, materials or products to which have been affixed a label, seal, symbol or other identifying mark of a nationally recognized testing laboratory, inspection agency or other organization concerned with product evaluation that maintains periodic inspection of the production of the above-labeled items and whose labeling indicates either that the equipment, material or product meets identified standards or has been tested and found suitable for a specified purpose.

LISTED. Equipment, materials, products or services included in a list published by an organization acceptable to the *code official* and concerned with evaluation of products or services that maintains periodic inspection of production of *listed* equipment or materials or periodic evaluation of services and whose listing states either that the equipment, material, product or service meets identified standards or has been tested and found suitable for a specified purpose.

LOW-VOLTAGE LIGHTING. Lighting equipment powered through a transformer such as a cable conductor, a rail conductor and track lighting.

MANUAL. Capable of being operated by personal intervention (see "Automatic").

NAMEPLATE HORSEPOWER. The nominal motor horsepower rating stamped on the motor nameplate.

PROPOSED DESIGN. A description of the proposed building used to estimate annual energy use for determining compliance based on total building performance.

READILY ACCESSIBLE. Capable of being reached quickly for operation, renewal or inspection without requiring those to whom ready access is requisite to climb over or remove obstacles or to resort to portable ladders or access equipment (see "*Accessible*").

REPAIR. The reconstruction or renewal of any part of an existing building.

RESIDENTIAL BUILDING. For this code, includes R-3 buildings, as well as R-2 and R-4 buildings three stories or less in height above grade.

ROOF ASSEMBLY. A system designed to provide weather protection and resistance to design loads. The system consists of a roof covering and roof deck or a single component serving as both the roof covering and the roof deck. A roof assembly includes the roof covering, underlayment, roof deck, insulation, vapor retarder and interior finish.

***R*-VALUE (THERMAL RESISTANCE).** The inverse of the time rate of heat flow through a body from one of its bounding surfaces to the other surface for a unit temperature difference between the two surfaces, under steady state conditions, per unit area ($h \cdot ft^2 \cdot °F/Btu$) [($m^2 \cdot K$)/W].

SCREW LAMP HOLDERS. A lamp base that requires a screw-in-type lamp, such as a compact-fluorescent, incandescent, or tungsten-halogen bulb.

SERVICE WATER HEATING. Supply of hot water for purposes other than comfort heating.

SKYLIGHT. Glass or other transparent or translucent glazing material installed at a slope of 15 degrees (0.26 rad) or more from vertical. Glazing material in skylights, including unit skylights, solariums, sunrooms, roofs and sloped walls is included in this definition.

SLEEPING UNIT. A room or space in which people sleep, which can also include permanent provisions for living, eating, and either sanitation or kitchen facilities but not both. Such rooms and spaces that are also part of a dwelling unit are not *sleeping units*.

SOLAR HEAT GAIN COEFFICIENT (SHGC). The ratio of the solar heat gain entering the space through the fenestration assembly to the incident solar radiation. Solar heat gain includes directly transmitted solar heat and absorbed solar radiation which is then reradiated, conducted or convected into the space.

STANDARD REFERENCE DESIGN. A version of the *proposed design* that meets the minimum requirements of this code and is used to determine the maximum annual energy use requirement for compliance based on total building performance.

STOREFRONT. A nonresidential system of doors and windows mulled as a composite fenestration structure that has been designed to resist heavy use. *Storefront* systems include, but are not limited to, exterior fenestration systems that span from the floor level or above to the ceiling of the same story on commercial buildings.

SUNROOM. A one-story structure attached to a dwelling with a glazing area in excess of 40 percent of the gross area of the structure's exterior walls and roof.

THERMAL ISOLATION. Physical and space conditioning separation from *conditioned space(s)*. The *conditioned space*(s) shall be controlled as separate zones for heating and cooling or conditioned by separate equipment.

THERMOSTAT. An automatic control device used to maintain temperature at a fixed or adjustable set point.

***U*-FACTOR (THERMAL TRANSMITTANCE).** The coefficient of heat transmission (air to air) through a building component or assembly, equal to the time rate of heat flow per unit area and unit temperature difference between the warm side and cold side air films ($Btu/h \cdot ft^2 \cdot °F$) [$W/(m^2 \cdot K)$].

VENTILATION. The natural or mechanical process of supplying conditioned or unconditioned air to, or removing such air from, any space.

VENTILATION AIR. That portion of supply air that comes from outside (outdoors) plus any recirculated air that has been treated to maintain the desired quality of air within a designated space.

ZONE. A space or group of spaces within a building with heating or cooling requirements that are sufficiently similar so that desired conditions can be maintained throughout using a single controlling device.

CHAPTER 3
GENERAL REQUIREMENTS

**SECTION 301
CLIMATE ZONES**

301.1 General. Climate *zones* from Figure 301.1 or Table 301.1 shall be used in determining the applicable requirements from Chapters 4 and 5. Locations not in Table 301.1 (outside the United States) shall be assigned a climate *zone* based on Section 301.3.

301.2 Warm humid counties. Warm humid counties are identified in Table 301.1 by an asterisk.

301.3 International climate zones. The climate *zone* for any location outside the United States shall be determined by applying Table 301.3(1) and then Table 301.3(2).

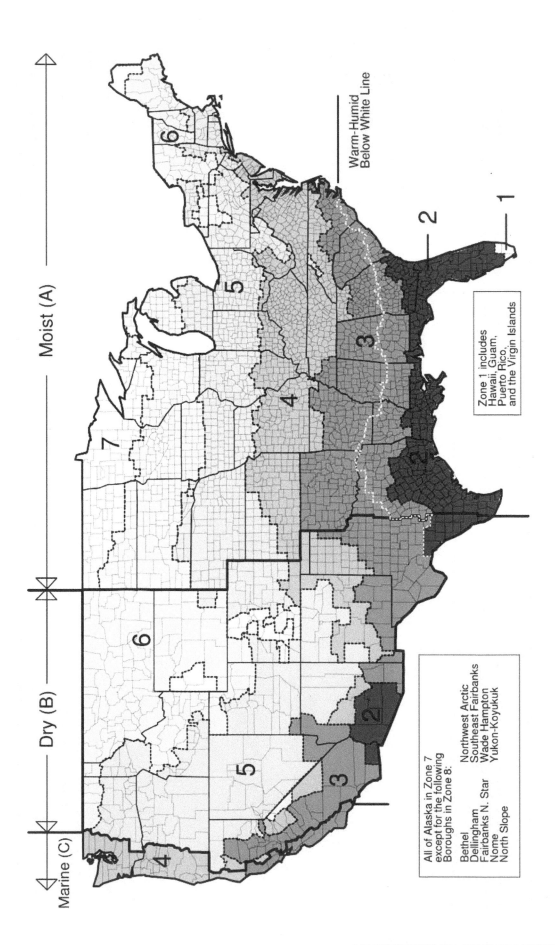

FIGURE 301.1
CLIMATE ZONES

Moist (A)

Dry (B)

Marine (C)

Warm-Humid
Below White Line

Zone 1 includes
Hawaii, Guam,
Puerto Rico,
and the Virgin Islands

All of Alaska in Zone 7
except for the following
Boroughs in Zone 8:

Bethel   Northwest Arctic
Dellingham   Southeast Fairbanks
Fairbanks N. Star   Wade Hampton
Nome   Yukon-Koyukuk
North Slope

TABLE 301.1
CLIMATE ZONES, MOISTURE REGIMES, AND WARM-HUMID DESIGNATIONS
BY STATE, COUNTY AND TERRITORY

Note: Table 301.1 in the 2006 edition has been replaced in its entirety. Margin lines are omitted for clarity.

Key: A – Moist, B – Dry, C – Marine. Absence of moisture designation indicates moisture regime is irrelevant. Asterisk (*) indicates a warm-humid location.

US STATES

ALABAMA
3A Autauga*
2A Baldwin*
3A Barbour*
3A Bibb
3A Blount
3A Bullock*
3A Butler*
3A Calhoun
3A Chambers
3A Cherokee
3A Chilton
3A Choctaw*
3A Clarke*
3A Clay
3A Cleburne
3A Coffee*
3A Colbert
3A Conecuh*
3A Coosa
3A Covington*
3A Crenshaw*
3A Cullman
3A Dale*
3A Dallas*
3A DeKalb
3A Elmore*
3A Escambia*
3A Etowah
3A Fayette
3A Franklin
3A Geneva*
3A Greene
3A Hale

3A Henry*
3A Houston*
3A Jackson
3A Jefferson
3A Lamar
3A Lauderdale
3A Lawrence
3A Lee
3A Limestone
3A Lowndes*
3A Macon*
3A Madison
3A Marengo*
3A Marion
3A Marshall
2A Mobile*
3A Monroe*
3A Montgomery*
3A Morgan
3A Perry*
3A Pickens
3A Pike*
3A Randolph
3A Russell*
3A Shelby
3A St. Clair
3A Sumter
3A Talladega
3A Tallapoosa
3A Tuscaloosa
3A Walker
3A Washington*
3A Wilcox*
3A Winston

ALASKA
7 Aleutians East
7 Aleutians West
7 Anchorage
8 Bethel
7 Bristol Bay
7 Denali
8 Dillingham
8 Fairbanks North Star
7 Haines
7 Juneau
7 Kenai Peninsula
7 Ketchikan Gateway
7 Kodiak Island
7 Lake and Peninsula
7 Matanuska-Susitna
8 Nome
8 North Slope
8 Northwest Arctic
7 Prince of Wales-Outer Ketchikan
7 Sitka
7 Skagway-Hoonah-Angoon
8 Southeast Fairbanks
7 Valdez-Cordova
8 Wade Hampton
7 Wrangell-Petersburg
7 Yakutat
8 Yukon-Koyukuk

ARIZONA
5B Apache
3B Cochise

5B Coconino
4B Gila
3B Graham
3B Greenlee
2B La Paz
2B Maricopa
3B Mohave
5B Navajo
2B Pima
2B Pinal
3B Santa Cruz
4B Yavapai
2B Yuma

ARKANSAS
3A Arkansas
3A Ashley
4A Baxter
4A Benton
4A Boone
3A Bradley
3A Calhoun
4A Carroll
3A Chicot
3A Clark
3A Clay
3A Cleburne
3A Cleveland
3A Columbia*
3A Conway
3A Craighead
3A Crawford
3A Crittenden
3A Cross
3A Dallas

3A Desha
3A Drew
3A Faulkner
3A Franklin
4A Fulton
3A Garland
3A Grant
3A Greene
3A Hempstead*
3A Hot Spring
3A Howard
3A Independence
4A Izard
3A Jackson
3A Jefferson
3A Johnson
3A Lafayette*
3A Lawrence
3A Lee
3A Lincoln
3A Little River*
3A Logan
3A Lonoke
4A Madison
4A Marion
3A Miller*
3A Mississippi
3A Monroe
3A Montgomery
3A Nevada
4A Newton
3A Ouachita
3A Perry
3A Phillips

(continued)

TABLE 301.1—continued
CLIMATE ZONES, MOISTURE REGIMES, AND WARM-HUMID DESIGNATIONS
BY STATE, COUNTY AND TERRITORY

3A Pike	3B Los Angeles	**COLORADO**	7 Mineral	2A Charlotte*
3A Poinsett	3B Madera	5B Adams	6B Moffat	2A Citrus*
3A Polk	3C Marin	6B Alamosa	5B Montezuma	2A Clay*
3A Pope	4B Mariposa	5B Arapahoe	5B Montrose	2A Collier*
3A Prairie	3C Mendocino	6B Archuleta	5B Morgan	2A Columbia*
3A Pulaski	3B Merced	4B Baca	4B Otero	2A DeSoto*
3A Randolph	5B Modoc	5B Bent	6B Ouray	2A Dixie*
3A Saline	6B Mono	5B Boulder	7 Park	2A Duval*
3A Scott	3C Monterey	6B Chaffee	5B Phillips	2A Escambia*
4A Searcy	3C Napa	5B Cheyenne	7 Pitkin	2A Flagler*
3A Sebastian	5B Nevada	7 Clear Creek	5B Prowers	2A Franklin*
3A Sevier*	3B Orange	6B Conejos	5B Pueblo	2A Gadsden*
3A Sharp	3B Placer	6B Costilla	6B Rio Blanco	2A Gilchrist*
3A St. Francis	5B Plumas	5B Crowley	7 Rio Grande	2A Glades*
4A Stone	3B Riverside	6B Custer	7 Routt	2A Gulf*
3A Union*	3B Sacramento	5B Delta	6B Saguache	2A Hamilton*
3A Van Buren	3C San Benito	5B Denver	7 San Juan	2A Hardee*
4A Washington	3B San Bernardino	6B Dolores	6B San Miguel	2A Hendry*
3A White		5B Douglas	5B Sedgwick	2A Hernando*
3A Woodruff	3B San Diego	6B Eagle	7 Summit	2A Highlands*
3A Yell	3C San Francisco	5B Elbert	5B Teller	2A Hillsborough*
	3B San Joaquin	5B El Paso	5B Washington	2A Holmes*
CALIFORNIA	3C San Luis Obispo	5B Fremont	5B Weld	2A Indian River*
3C Alameda	3C San Mateo	5B Garfield	5B Yuma	2A Jackson*
6B Alpine	3C Santa Barbara	5B Gilpin		2A Jefferson*
4B Amador	3C Santa Clara	7 Grand	**CONNECTICUT**	2A Lafayette*
3B Butte	3C Santa Cruz	7 Gunnison	5A (all)	2A Lake*
4B Calaveras	3B Shasta	7 Hinsdale		2A Lee*
3B Colusa	5B Sierra	5B Huerfano	**DELAWARE**	2A Leon*
3B Contra Costa	5B Siskiyou	7 Jackson	4A (all)	2A Levy*
4C Del Norte	3B Solano	5B Jefferson		2A Liberty*
4B El Dorado	3C Sonoma	5B Kiowa	**DISTRICT OF COLUMBIA**	2A Madison*
3B Fresno	3B Stanislaus	5B Kit Carson		2A Manatee*
3B Glenn	3B Sutter	7 Lake	4A (all)	2A Marion*
4C Humboldt	3B Tehama	5B La Plata		2A Martin*
2B Imperial	4B Trinity	5B Larimer	**FLORIDA**	1A Miami-Dade*
4B Inyo	3B Tulare	4B Las Animas	2A Alachua*	1A Monroe*
3B Kern	4B Tuolumne	5B Lincoln	2A Baker*	2A Nassau*
3B Kings	3C Ventura	5B Logan	2A Bay*	2A Okaloosa*
4B Lake	3B Yolo	5B Mesa	2A Bradford*	2A Okeechobee*
5B Lassen	3B Yuba		2A Brevard*	
			1A Broward*	
			2A Calhoun*	

(continued)

2009 INTERNATIONAL ENERGY CONSERVATION CODE®

TABLE 301.1—continued
CLIMATE ZONES, MOISTURE REGIMES, AND WARM-HUMID DESIGNATIONS
BY STATE, COUNTY AND TERRITORY

2A Orange*	2A Camden*	4A Gilmer	3A Monroe	3A Twiggs*
2A Osceola*	3A Candler*	3A Glascock	3A Montgomery*	4A Union
2A Palm Beach*	3A Carroll	2A Glynn*	3A Morgan	3A Upson
2A Pasco*	4A Catoosa	4A Gordon	4A Murray	4A Walker
2A Pinellas*	2A Charlton*	2A Grady*	3A Muscogee	3A Walton
2A Polk*	2A Chatham*	3A Greene	3A Newton	2A Ware*
2A Putnam*	3A Chattahoochee*	3A Gwinnett	3A Oconee	3A Warren
2A Santa Rosa*	4A Chattooga	4A Habersham	3A Oglethorpe	3A Washington
2A Sarasota*	3A Cherokee	4A Hall	3A Paulding	2A Wayne*
2A Seminole*	3A Clarke	3A Hancock	3A Peach*	3A Webster*
2A St. Johns*	3A Clay*	3A Haralson	4A Pickens	3A Wheeler*
2A St. Lucie*	3A Clayton	3A Harris	2A Pierce*	4A White
2A Sumter*	2A Clinch*	3A Hart	3A Pike	4A Whitfield
2A Suwannee*	3A Cobb	3A Heard	3A Polk	3A Wilcox*
2A Taylor*	3A Coffee*	3A Henry	3A Pulaski*	3A Wilkes
2A Union*	2A Colquitt*	3A Houston*	3A Putnam	3A Wilkinson
2A Volusia*	3A Columbia	3A Irwin*	3A Quitman*	3A Worth*
2A Wakulla*	2A Cook*	3A Jackson	4A Rabun	
2A Walton*	3A Coweta	3A Jasper	3A Randolph*	**HAWAII**
2A Washington*	3A Crawford	2A Jeff Davis*	3A Richmond	1A (all)*
	3A Crisp*	3A Jefferson	3A Rockdale	
GEORGIA	4A Dade	3A Jenkins*	3A Schley*	**IDAHO**
2A Appling*	4A Dawson	3A Johnson*	3A Screven*	5B Ada
2A Atkinson*	2A Decatur*	3A Jones	2A Seminole*	6B Adams
2A Bacon*	3A DeKalb	3A Lamar	3A Spalding	6B Bannock
2A Baker*	3A Dodge*	2A Lanier*	4A Stephens	6B Bear Lake
3A Baldwin	3A Dooly*	3A Laurens*	3A Stewart*	5B Benewah
4A Banks	3A Dougherty*	3A Lee*	3A Sumter*	6B Bingham
3A Barrow	3A Douglas	2A Liberty*	3A Talbot	6B Blaine
3A Bartow	3A Early*	3A Lincoln	3A Taliaferro	6B Boise
3A Ben Hill*	2A Echols*	2A Long*	2A Tattnall*	6B Bonner
2A Berrien*	2A Effingham*	2A Lowndes*	3A Taylor*	6B Bonneville
3A Bibb	3A Elbert	4A Lumpkin	3A Telfair*	6B Boundary
3A Bleckley*	3A Emanuel*	3A Macon*	3A Terrell*	6B Butte
2A Brantley*	2A Evans*	3A Madison	2A Thomas*	6B Camas
2A Brooks*	4A Fannin	3A Marion*	3A Tift*	5B Canyon
2A Bryan*	3A Fayette	3A McDuffie	2A Toombs*	6B Caribou
3A Bulloch*	4A Floyd	2A McIntosh*	4A Towns	5B Cassia
3A Burke	3A Forsyth	3A Meriwether	3A Treutlen*	6B Clark
3A Butts	4A Franklin	2A Miller*	3A Troup	5B Clearwater
3A Calhoun*	3A Fulton	2A Mitchell*	3A Turner*	6B Custer
				5B Elmore

(continued)

TABLE 301.1—continued
CLIMATE ZONES, MOISTURE REGIMES, AND WARM-HUMID DESIGNATIONS
BY STATE, COUNTY AND TERRITORY

6B Franklin	5A Cook	4A Macoupin	4A Wayne	5A Henry
6B Fremont	4A Crawford	4A Madison	4A White	5A Howard
5B Gem	5A Cumberland	4A Marion	5A Whiteside	5A Huntington
5B Gooding	5A DeKalb	5A Marshall	5A Will	4A Jackson
5B Idaho	5A De Witt	5A Mason	4A Williamson	5A Jasper
6B Jefferson	5A Douglas	4A Massac	5A Winnebago	5A Jay
5B Jerome	5A DuPage	5A McDonough	5A Woodford	4A Jefferson
5B Kootenai	5A Edgar	5A McHenry		4A Jennings
5B Latah	4A Edwards	5A McLean	**INDIANA**	5A Johnson
6B Lemhi	4A Effingham	5A Menard	5A Adams	4A Knox
5B Lewis	4A Fayette	5A Mercer	5A Allen	5A Kosciusko
5B Lincoln	5A Ford	4A Monroe	5A Bartholomew	5A Lagrange
6B Madison	4A Franklin	4A Montgomery	5A Benton	5A Lake
5B Minidoka	5A Fulton	5A Morgan	5A Blackford	5A La Porte
5B Nez Perce	4A Gallatin	5A Moultrie	5A Boone	4A Lawrence
6B Oneida	5A Greene	5A Ogle	4A Brown	5A Madison
5B Owyhee	5A Grundy	5A Peoria	5A Carroll	5A Marion
5B Payette	4A Hamilton	4A Perry	5A Cass	5A Marshall
5B Power	5A Hancock	5A Piatt	4A Clark	4A Martin
5B Shoshone	4A Hardin	5A Pike	5A Clay	5A Miami
6B Teton	5A Henderson	4A Pope	5A Clinton	4A Monroe
5B Twin Falls	5A Henry	4A Pulaski	4A Crawford	5A Montgomery
6B Valley	5A Iroquois	5A Putnam	4A Daviess	5A Morgan
5B Washington	4A Jackson	4A Randolph	4A Dearborn	5A Newton
	4A Jasper	4A Richland	5A Decatur	5A Noble
ILLINOIS	4A Jefferson	5A Rock Island	5A De Kalb	4A Ohio
5A Adams	5A Jersey	4A Saline	5A Delaware	4A Orange
4A Alexander	5A Jo Daviess	5A Sangamon	4A Dubois	5A Owen
4A Bond	4A Johnson	5A Schuyler	5A Elkhart	5A Parke
5A Boone	5A Kane	5A Scott	5A Fayette	4A Perry
5A Brown	5A Kankakee	4A Shelby	4A Floyd	4A Pike
5A Bureau	5A Kendall	5A Stark	5A Fountain	5A Porter
5A Calhoun	5A Knox	4A St. Clair	5A Franklin	4A Posey
5A Carroll	5A Lake	5A Stephenson	5A Fulton	5A Pulaski
5A Cass	5A La Salle	5A Tazewell	4A Gibson	5A Putnam
5A Champaign	4A Lawrence	4A Union	5A Grant	5A Randolph
4A Christian	5A Lee	5A Vermilion	4A Greene	4A Ripley
5A Clark	5A Livingston	4A Wabash	5A Hamilton	5A Rush
4A Clay	5A Logan	5A Warren	5A Hancock	4A Scott
4A Clinton	5A Macon	4A Washington	4A Harrison	5A Shelby
5A Coles			5A Hendricks	

(continued)

TABLE 301.1—continued
CLIMATE ZONES, MOISTURE REGIMES, AND WARM-HUMID DESIGNATIONS
BY STATE, COUNTY AND TERRITORY

4A Spencer	5A Clarke	6A Lyon	**KANSAS**	4A Harvey
5A Starke	6A Clay	5A Madison	4A Allen	4A Haskell
5A Steuben	6A Clayton	5A Mahaska	4A Anderson	4A Hodgeman
5A St. Joseph	5A Clinton	5A Marion	4A Atchison	4A Jackson
4A Sullivan	5A Crawford	5A Marshall	4A Barber	4A Jefferson
4A Switzerland	5A Dallas	5A Mills	4A Barton	5A Jewell
5A Tippecanoe	5A Davis	6A Mitchell	4A Bourbon	4A Johnson
5A Tipton	5A Decatur	5A Monona	4A Brown	4A Kearny
5A Union	6A Delaware	5A Monroe	4A Butler	4A Kingman
4A Vanderburgh	5A Des Moines	5A Montgomery	4A Chase	4A Kiowa
5A Vermillion	6A Dickinson	5A Muscatine	4A Chautauqua	4A Labette
5A Vigo	5A Dubuque	6A O'Brien	4A Cherokee	5A Lane
5A Wabash	6A Emmet	6A Osceola	5A Cheyenne	4A Leavenworth
5A Warren	6A Fayette	5A Page	4A Clark	4A Lincoln
4A Warrick	6A Floyd	6A Palo Alto	4A Clay	4A Linn
4A Washington	6A Franklin	6A Plymouth	5A Cloud	5A Logan
5A Wayne	5A Fremont	6A Pocahontas	4A Coffey	4A Lyon
5A Wells	5A Greene	5A Polk	4A Comanche	4A Marion
5A White	6A Grundy	5A Pottawattamie	4A Cowley	4A Marshall
5A Whitley	5A Guthrie	5A Poweshiek	4A Crawford	4A McPherson
IOWA	6A Hamilton	5A Ringgold	5A Decatur	4A Meade
5A Adair	6A Hancock	6A Sac	4A Dickinson	4A Miami
5A Adams	6A Hardin	5A Scott	4A Doniphan	5A Mitchell
6A Allamakee	5A Harrison	5A Shelby	4A Douglas	4A Montgomery
5A Appanoose	5A Henry	6A Sioux	4A Edwards	4A Morris
5A Audubon	6A Howard	5A Story	4A Elk	4A Morton
5A Benton	6A Humboldt	5A Tama	5A Ellis	4A Nemaha
6A Black Hawk	6A Ida	5A Taylor	4A Ellsworth	4A Neosho
5A Boone	5A Iowa	5A Union	4A Finney	5A Ness
6A Bremer	5A Jackson	5A Van Buren	4A Ford	5A Norton
6A Buchanan	5A Jasper	5A Wapello	4A Franklin	4A Osage
6A Buena Vista	5A Jefferson	5A Warren	4A Geary	5A Osborne
6A Butler	5A Johnson	5A Washington	5A Gove	4A Ottawa
6A Calhoun	5A Jones	5A Wayne	5A Graham	4A Pawnee
5A Carroll	5A Keokuk	6A Webster	4A Grant	5A Phillips
5A Cass	6A Kossuth	6A Winnebago	4A Gray	4A Pottawatomie
5A Cedar	5A Lee	6A Winneshiek	5A Greeley	4A Pratt
6A Cerro Gordo	5A Linn	5A Woodbury	4A Greenwood	5A Rawlins
6A Cherokee	5A Louisa	6A Worth	5A Hamilton	4A Reno
6A Chickasaw	5A Lucas	6A Wright	4A Harper	5A Republic

(continued)

TABLE 301.1—continued
CLIMATE ZONES, MOISTURE REGIMES, AND WARM-HUMID DESIGNATIONS
BY STATE, COUNTY AND TERRITORY

4A Rice	2A Cameron*	2A St. Mary*	4A Cecil	6A Crawford
4A Riley	3A Catahoula*	2A St. Tammany*	4A Charles	6A Delta
5A Rooks	3A Claiborne*	2A Tangipahoa*	4A Dorchester	6A Dickinson
4A Rush	3A Concordia*	3A Tensas*	4A Frederick	5A Eaton
4A Russell	3A De Soto*	2A Terrebonne*	5A Garrett	6A Emmet
4A Saline	2A East Baton Rouge*	3A Union*	4A Harford	5A Genesee
5A Scott	3A East Carroll	2A Vermilion*	4A Howard	6A Gladwin
4A Sedgwick	2A East Feliciana*	3A Vernon*	4A Kent	7 Gogebic
4A Seward	2A Evangeline*	2A Washington*	4A Montgomery	6A Grand Traverse
4A Shawnee	3A Franklin*	3A Webster*	4A Prince George's	5A Gratiot
5A Sheridan	3A Grant*	2A West Baton Rouge*	4A Queen Anne's	5A Hillsdale
5A Sherman	2A Iberia*		4A Somerset	7 Houghton
5A Smith	2A Iberville*	3A West Carroll	4A St. Mary's	6A Huron
4A Stafford	3A Jackson*	2A West Feliciana*	4A Talbot	5A Ingham
4A Stanton	2A Jefferson*	3A Winn*	4A Washington	5A Ionia
4A Stevens	2A Jefferson Davis*	**MAINE**	4A Wicomico	6A Iosco
4A Sumner	2A Lafayette*	6A Androscoggin	4A Worcester	7 Iron
5A Thomas	2A Lafourche*	7 Aroostook	**MASSACHUSETTS**	6A Isabella
5A Trego	3A La Salle*	6A Cumberland	5A (all)	5A Jackson
4A Wabaunsee	3A Lincoln*	6A Franklin	**MICHIGAN**	5A Kalamazoo
5A Wallace	2A Livingston*	6A Hancock	6A Alcona	6A Kalkaska
4A Washington	3A Madison*	6A Kennebec	6A Alger	5A Kent
5A Wichita	3A Morehouse	6A Knox	5A Allegan	7 Keweenaw
4A Wilson	3A Natchitoches*	6A Lincoln	6A Alpena	6A Lake
4A Woodson	2A Orleans*	6A Oxford	6A Antrim	5A Lapeer
4A Wyandotte	3A Ouachita*	6A Penobscot	6A Arenac	6A Leelanau
KENTUCKY	2A Plaquemines*	6A Piscataquis	7 Baraga	5A Lenawee
4A (all)	2A Pointe Coupee*	6A Sagadahoc	5A Barry	5A Livingston
LOUISIANA	2A Rapides*	6A Somerset	5A Bay	7 Luce
2A Acadia*	3A Red River*	6A Waldo	6A Benzie	7 Mackinac
2A Allen*	3A Richland*	6A Washington	5A Berrien	5A Macomb
2A Ascension*	3A Sabine*	6A York	5A Branch	6A Manistee
2A Assumption*	2A St. Bernard*	**MARYLAND**	5A Calhoun	6A Marquette
2A Avoyelles*	2A St. Charles*	4A Allegany	5A Cass	6A Mason
2A Beauregard*	2A St. Helena*	4A Anne Arundel	6A Charlevoix	6A Mecosta
3A Bienville*	2A St. James*	4A Baltimore	6A Cheboygan	6A Menominee
3A Bossier*	2A St. John the Baptist*	4A Baltimore (city)	7 Chippewa	5A Midland
3A Caddo*		4A Calvert	6A Clare	6A Missaukee
2A Calcasieu*	2A St. Landry*	4A Caroline	5A Clinton	5A Monroe
3A Caldwell*	2A St. Martin*	4A Carroll		5A Montcalm

(continued)

TABLE 301.1—continued
CLIMATE ZONES, MOISTURE REGIMES, AND WARM-HUMID DESIGNATIONS
BY STATE, COUNTY AND TERRITORY

6A Montmorency
5A Muskegon
6A Newaygo
5A Oakland
6A Oceana
6A Ogemaw
7 Ontonagon
6A Osceola
6A Oscoda
6A Otsego
5A Ottawa
6A Presque Isle
6A Roscommon
5A Saginaw
6A Sanilac
7 Schoolcraft
5A Shiawassee
5A St. Clair
5A St. Joseph
5A Tuscola
5A Van Buren
5A Washtenaw
5A Wayne
6A Wexford

MINNESOTA

7 Aitkin
6A Anoka
7 Becker
7 Beltrami
6A Benton
6A Big Stone
6A Blue Earth
6A Brown
7 Carlton
6A Carver
7 Cass
6A Chippewa
6A Chisago
7 Clay
7 Clearwater

7 Cook
6A Cottonwood
7 Crow Wing
6A Dakota
6A Dodge
6A Douglas
6A Faribault
6A Fillmore
6A Freeborn
6A Goodhue
7 Grant
6A Hennepin
6A Houston
7 Hubbard
6A Isanti
7 Itasca
6A Jackson
7 Kanabec
6A Kandiyohi
7 Kittson
7 Koochiching
6A Lac qui Parle
7 Lake
7 Lake of the Woods
6A Le Sueur
6A Lincoln
6A Lyon
7 Mahnomen
7 Marshall
6A Martin
6A McLeod
6A Meeker
7 Mille Lacs
6A Morrison
6A Mower
6A Murray
6A Nicollet
6A Nobles
7 Norman
6A Olmsted
7 Otter Tail

7 Pennington
7 Pine
6A Pipestone
7 Polk
6A Pope
6A Ramsey
7 Red Lake
6A Redwood
6A Renville
6A Rice
6A Rock
7 Roseau
6A Scott
6A Sherburne
6A Sibley
6A Stearns
6A Steele
6A Stevens
7 St. Louis
6A Swift
6A Todd
6A Traverse
6A Wabasha
7 Wadena
6A Waseca
6A Washington
6A Watonwan
7 Wilkin
6A Winona
6A Wright
6A Yellow
 Medicine

MISSISSIPPI

3A Adams*
3A Alcorn
3A Amite*
3A Attala
3A Benton
3A Bolivar
3A Calhoun

3A Carroll
3A Chickasaw
3A Choctaw
3A Claiborne*
3A Clarke
3A Clay
3A Coahoma
3A Copiah*
3A Covington*
3A DeSoto
3A Forrest*
3A Franklin*
3A George*
3A Greene*
3A Grenada
2A Hancock*
2A Harrison*
3A Hinds*
3A Holmes
3A Humphreys
3A Issaquena
3A Itawamba
2A Jackson*
3A Jasper
3A Jefferson*
3A Jefferson Davis*
3A Jones*
3A Kemper
3A Lafayette
3A Lamar*
3A Lauderdale
3A Lawrence*
3A Leake
3A Lee
3A Leflore
3A Lincoln*
3A Lowndes
3A Madison
3A Marion*
3A Marshall
3A Monroe

3A Montgomery
3A Neshoba
3A Newton
3A Noxubee
3A Oktibbeha
3A Panola
2A Pearl River*
3A Perry*
3A Pike*
3A Pontotoc
3A Prentiss
3A Quitman
3A Rankin*
3A Scott
3A Sharkey
3A Simpson*
3A Smith*
2A Stone*
3A Sunflower
3A Tallahatchie
3A Tate
3A Tippah
3A Tishomingo
3A Tunica
3A Union
3A Walthall*
3A Warren*
3A Washington
3A Wayne*
3A Webster
3A Wilkinson*
3A Winston
3A Yalobusha
3A Yazoo

MISSOURI

5A Adair
5A Andrew
5A Atchison
4A Audrain
4A Barry

(continued)

TABLE 301.1—continued
CLIMATE ZONES, MOISTURE REGIMES, AND WARM-HUMID DESIGNATIONS
BY STATE, COUNTY AND TERRITORY

4A Barton

4A Bates

4A Benton

4A Bollinger

4A Boone

5A Buchanan

4A Butler

5A Caldwell

4A Callaway

4A Camden

4A Cape Girardeau

4A Carroll

4A Carter

4A Cass

4A Cedar

5A Chariton

4A Christian

5A Clark

4A Clay

5A Clinton

4A Cole

4A Cooper

4A Crawford

4A Dade

4A Dallas

5A Daviess

5A DeKalb

4A Dent

4A Douglas

4A Dunklin

4A Franklin

4A Gasconade

5A Gentry

4A Greene

5A Grundy

5A Harrison

4A Henry

4A Hickory

5A Holt

4A Howard

4A Howell

4A Iron

4A Jackson

4A Jasper

4A Jefferson

4A Johnson

5A Knox

4A Laclede

4A Lafayette

4A Lawrence

5A Lewis

4A Lincoln

5A Linn

5A Livingston

5A Macon

4A Madison

4A Maries

5A Marion

4A McDonald

5A Mercer

4A Miller

4A Mississippi

4A Moniteau

4A Monroe

4A Montgomery

4A Morgan

4A New Madrid

4A Newton

5A Nodaway

4A Oregon

4A Osage

4A Ozark

4A Pemiscot

4A Perry

4A Pettis

4A Phelps

5A Pike

4A Platte

4A Polk

4A Pulaski

5A Putnam

5A Ralls

4A Randolph

4A Ray

4A Reynolds

4A Ripley

4A Saline

5A Schuyler

5A Scotland

4A Scott

4A Shannon

5A Shelby

4A St. Charles

4A St. Clair

4A Ste. Genevieve

4A St. Francois

4A St. Louis

4A St. Louis (city)

4A Stoddard

4A Stone

5A Sullivan

4A Taney

4A Texas

4A Vernon

4A Warren

4A Washington

4A Wayne

4A Webster

5A Worth

4A Wright

MONTANA

6B (all)

NEBRASKA

5A (all)

NEVADA

5B Carson City (city)

5B Churchill

3B Clark

5B Douglas

5B Elko

5B Esmeralda

5B Eureka

5B Humboldt

5B Lander

5B Lincoln

5B Lyon

5B Mineral

5B Nye

5B Pershing

5B Storey

5B Washoe

5B White Pine

NEW HAMPSHIRE

6A Belknap

6A Carroll

5A Cheshire

6A Coos

6A Grafton

5A Hillsborough

6A Merrimack

5A Rockingham

5A Strafford

6A Sullivan

NEW JERSEY

4A Atlantic

5A Bergen

4A Burlington

4A Camden

4A Cape May

4A Cumberland

4A Essex

4A Gloucester

4A Hudson

5A Hunterdon

5A Mercer

4A Middlesex

4A Monmouth

5A Morris

4A Ocean

5A Passaic

4A Salem

5A Somerset

5A Sussex

4A Union

5A Warren

NEW MEXICO

4B Bernalillo

5B Catron

3B Chaves

4B Cibola

5B Colfax

4B Curry

4B DeBaca

3B Dona Ana

3B Eddy

4B Grant

4B Guadalupe

5B Harding

3B Hidalgo

3B Lea

4B Lincoln

5B Los Alamos

3B Luna

5B McKinley

5B Mora

3B Otero

4B Quay

5B Rio Arriba

4B Roosevelt

5B Sandoval

5B San Juan

5B San Miguel

5B Santa Fe

4B Sierra

4B Socorro

5B Taos

5B Torrance

4B Union

4B Valencia

(continued)

TABLE 301.1—continued
CLIMATE ZONES, MOISTURE REGIMES, AND WARM-HUMID DESIGNATIONS
BY STATE, COUNTY AND TERRITORY

NEW YORK

5A Albany
6A Allegany
4A Bronx
6A Broome
6A Cattaraugus
5A Cayuga
5A Chautauqua
5A Chemung
6A Chenango
6A Clinton
5A Columbia
5A Cortland
6A Delaware
5A Dutchess
5A Erie
6A Essex
6A Franklin
6A Fulton
5A Genesee
5A Greene
6A Hamilton
6A Herkimer
6A Jefferson
4A Kings
6A Lewis
5A Livingston
6A Madison
5A Monroe
6A Montgomery
4A Nassau
4A New York
5A Niagara
6A Oneida
5A Onondaga
5A Ontario
5A Orange
5A Orleans
5A Oswego
6A Otsego

5A Putnam
4A Queens
5A Rensselaer
4A Richmond
5A Rockland
5A Saratoga
5A Schenectady
6A Schoharie
6A Schuyler
5A Seneca
6A Steuben
6A St. Lawrence
4A Suffolk
6A Sullivan
5A Tioga
6A Tompkins
6A Ulster
6A Warren
5A Washington
5A Wayne
4A Westchester
6A Wyoming
5A Yates

NORTH CAROLINA

4A Alamance
4A Alexander
5A Alleghany
3A Anson
5A Ashe
5A Avery
3A Beaufort
4A Bertie
3A Bladen
3A Brunswick*
4A Buncombe
4A Burke
3A Cabarrus
4A Caldwell
3A Camden

3A Carteret*
4A Caswell
4A Catawba
4A Chatham
4A Cherokee
3A Chowan
4A Clay
4A Cleveland
3A Columbus*
3A Craven
3A Cumberland
3A Currituck
3A Dare
3A Davidson
4A Davie
3A Duplin
4A Durham
3A Edgecombe
4A Forsyth
4A Franklin
3A Gaston
4A Gates
4A Graham
4A Granville
3A Greene
4A Guilford
4A Halifax
4A Harnett
4A Haywood
4A Henderson
4A Hertford
3A Hoke
3A Hyde
4A Iredell
4A Jackson
3A Johnston
3A Jones
4A Lee
3A Lenoir
4A Lincoln
4A Macon

4A Madison
3A Martin
4A McDowell
3A Mecklenburg
5A Mitchell
3A Montgomery
3A Moore
4A Nash
3A New Hanover*
4A Northampton
3A Onslow*
4A Orange
3A Pamlico
3A Pasquotank
3A Pender*
3A Perquimans
4A Person
3A Pitt
4A Polk
3A Randolph
3A Richmond
3A Robeson
4A Rockingham
3A Rowan
4A Rutherford
3A Sampson
3A Scotland
3A Stanly
4A Stokes
4A Surry
4A Swain
4A Transylvania
3A Tyrrell
3A Union
4A Vance
4A Wake
4A Warren
3A Washington
5A Watauga
3A Wayne
4A Wilkes

3A Wilson
4A Yadkin
5A Yancey

NORTH DAKOTA

6A Adams
7 Barnes
7 Benson
6A Billings
7 Bottineau
6A Bowman
7 Burke
6A Burleigh
7 Cass
7 Cavalier
6A Dickey
7 Divide
6A Dunn
7 Eddy
6A Emmons
7 Foster
6A Golden Valley
7 Grand Forks
6A Grant
7 Griggs
6A Hettinger
7 Kidder
6A LaMoure
6A Logan
7 McHenry
6A McIntosh
6A McKenzie
7 McLean
6A Mercer
6A Morton
7 Mountrail
7 Nelson
6A Oliver
7 Pembina
7 Pierce
7 Ramsey

(continued)

6A Ransom	5A Fairfield	5A Perry	3A Coal	3A Okmulgee
7 Renville	5A Fayette	5A Pickaway	3A Comanche	3A Osage
6A Richland	5A Franklin	4A Pike	3A Cotton	3A Ottawa
7 Rolette	5A Fulton	5A Portage	3A Craig	3A Pawnee
6A Sargent	4A Gallia	5A Preble	3A Creek	3A Payne
7 Sheridan	5A Geauga	5A Putnam	3A Custer	3A Pittsburg
6A Sioux	5A Greene	5A Richland	3A Delaware	3A Pontotoc
6A Slope	5A Guernsey	5A Ross	3A Dewey	3A Pottawatomie
6A Stark	4A Hamilton	5A Sandusky	3A Ellis	3A Pushmataha
7 Steele	5A Hancock	4A Scioto	3A Garfield	3A Roger Mills
7 Stutsman	5A Hardin	5A Seneca	3A Garvin	3A Rogers
7 Towner	5A Harrison	5A Shelby	3A Grady	3A Seminole
7 Traill	5A Henry	5A Stark	3A Grant	3A Sequoyah
7 Walsh	5A Highland	5A Summit	3A Greer	3A Stephens
7 Ward	5A Hocking	5A Trumbull	3A Harmon	4B Texas
7 Wells	5A Holmes	5A Tuscarawas	3A Harper	3A Tillman
7 Williams	5A Huron	5A Union	3A Haskell	3A Tulsa
	5A Jackson	5A Van Wert	3A Hughes	3A Wagoner
OHIO	5A Jefferson	5A Vinton	3A Jackson	3A Washington
4A Adams	5A Knox	5A Warren	3A Jefferson	3A Washita
5A Allen	5A Lake	4A Washington	3A Johnston	3A Woods
5A Ashland	4A Lawrence	5A Wayne	3A Kay	3A Woodward
5A Ashtabula	5A Licking	5A Williams	3A Kingfisher	
5A Athens	5A Logan	5A Wood	3A Kiowa	**OREGON**
5A Auglaize	5A Lorain	5A Wyandot	3A Latimer	5B Baker
5A Belmont	5A Lucas		3A Le Flore	4C Benton
4A Brown	5A Madison	**OKLAHOMA**	3A Lincoln	4C Clackamas
5A Butler	5A Mahoning	3A Adair	3A Logan	4C Clatsop
5A Carroll	5A Marion	3A Alfalfa	3A Love	4C Columbia
5A Champaign	5A Medina	3A Atoka	3A Major	4C Coos
5A Clark	5A Meigs	4B Beaver	3A Marshall	5B Crook
4A Clermont	5A Mercer	3A Beckham	3A Mayes	4C Curry
5A Clinton	5A Miami	3A Blaine	3A McClain	5B Deschutes
5A Columbiana	5A Monroe	3A Bryan	3A McCurtain	4C Douglas
5A Coshocton	5A Montgomery	3A Caddo	3A McIntosh	5B Gilliam
5A Crawford	5A Morgan	3A Canadian	3A Murray	5B Grant
5A Cuyahoga	5A Morrow	3A Carter	3A Muskogee	5B Harney
5A Darke	5A Muskingum	3A Cherokee	3A Noble	5B Hood River
5A Defiance	5A Noble	3A Choctaw	3A Nowata	4C Jackson
5A Delaware	5A Ottawa	4B Cimarron	3A Okfuskee	5B Jefferson
5A Erie	5A Paulding	3A Cleveland	3A Oklahoma	4C Josephine

(continued)

TABLE 301.1—continued
CLIMATE ZONES, MOISTURE REGIMES, AND WARM-HUMID DESIGNATIONS
BY STATE, COUNTY AND TERRITORY

5B Klamath	5A Cumberland	5A Warren	3A Lee	6A Faulk
5B Lake	5A Dauphin	5A Washington	3A Lexington	6A Grant
4C Lane	4A Delaware	6A Wayne	3A Marion	5A Gregory
4C Lincoln	6A Elk	5A Westmoreland	3A Marlboro	6A Haakon
4C Linn	5A Erie	5A Wyoming	3A McCormick	6A Hamlin
5B Malheur	5A Fayette	4A York	3A Newberry	6A Hand
4C Marion	5A Forest		3A Oconee	6A Hanson
5B Morrow	5A Franklin	**RHODE ISLAND**	3A Orangeburg	6A Harding
4C Multnomah	5A Fulton	5A (all)	3A Pickens	6A Hughes
4C Polk	5A Greene		3A Richland	5A Hutchinson
5B Sherman	5A Huntingdon	**SOUTH CAROLINA**	3A Saluda	6A Hyde
4C Tillamook	5A Indiana	3A Abbeville	3A Spartanburg	5A Jackson
5B Umatilla	5A Jefferson	3A Aiken	3A Sumter	6A Jerauld
5B Union	5A Juniata	3A Allendale*	3A Union	6A Jones
5B Wallowa	5A Lackawanna	3A Anderson	3A Williamsburg	6A Kingsbury
5B Wasco	5A Lancaster	3A Bamberg*	3A York	6A Lake
4C Washington	5A Lawrence	3A Barnwell*		6A Lawrence
5B Wheeler	5A Lebanon	3A Beaufort*	**SOUTH DAKOTA**	6A Lincoln
4C Yamhill	5A Lehigh	3A Berkeley*	6A Aurora	6A Lyman
	5A Luzerne	3A Calhoun	6A Beadle	6A Marshall
PENNSYLVANIA	5A Lycoming	3A Charleston*	5A Bennett	6A McCook
5A Adams	6A McKean	3A Cherokee	5A Bon Homme	6A McPherson
5A Allegheny	5A Mercer	3A Chester	6A Brookings	6A Meade
5A Armstrong	5A Mifflin	3A Chesterfield	6A Brown	5A Mellette
5A Beaver	5A Monroe	3A Clarendon	6A Brule	6A Miner
5A Bedford	4A Montgomery	3A Colleton*	6A Buffalo	6A Minnehaha
5A Berks	5A Montour	3A Darlington	6A Butte	6A Moody
5A Blair	5A Northampton	3A Dillon	6A Campbell	6A Pennington
5A Bradford	5A Northumberland	3A Dorchester*	5A Charles Mix	6A Perkins
4A Bucks	5A Perry	3A Edgefield	6A Clark	6A Potter
5A Butler	4A Philadelphia	3A Fairfield	5A Clay	6A Roberts
5A Cambria	5A Pike	3A Florence	6A Codington	6A Sanborn
6A Cameron	6A Potter	3A Georgetown*	6A Corson	6A Shannon
5A Carbon	5A Schuylkill	3A Greenville	6A Custer	6A Spink
5A Centre	5A Snyder	3A Greenwood	6A Davison	6A Stanley
4A Chester	5A Somerset	3A Hampton*	6A Day	6A Sully
5A Clarion	5A Sullivan	3A Horry*	6A Deuel	5A Todd
6A Clearfield	6A Susquehanna	3A Jasper*	6A Dewey	5A Tripp
5A Clinton	6A Tioga	3A Kershaw	5A Douglas	6A Turner
5A Columbia	5A Union	3A Lancaster	6A Edmunds	5A Union
5A Crawford	5A Venango	3A Laurens	6A Fall River	6A Walworth

(continued)

TABLE 301.1—continued
CLIMATE ZONES, MOISTURE REGIMES, AND WARM-HUMID DESIGNATIONS
BY STATE, COUNTY AND TERRITORY

5A Yankton	3A Haywood	3A Shelby	4B Briscoe	2B Dimmit*
6A Ziebach	3A Henderson	4A Smith	2A Brooks*	4B Donley
	4A Henry	4A Stewart	3A Brown*	2A Duval*
TENNESSEE	4A Hickman	4A Sullivan	2A Burleson*	3A Eastland
4A Anderson	4A Houston	4A Sumner	3A Burnet*	3B Ector
4A Bedford	4A Humphreys	3A Tipton	2A Caldwell*	2B Edwards*
4A Benton	4A Jackson	4A Trousdale	2A Calhoun*	3A Ellis*
4A Bledsoe	4A Jefferson	4A Unicoi	3B Callahan	3B El Paso
4A Blount	4A Johnson	4A Union	2A Cameron*	3A Erath*
4A Bradley	4A Knox	4A Van Buren	3A Camp*	2A Falls*
4A Campbell	3A Lake	4A Warren	4B Carson	3A Fannin
4A Cannon	3A Lauderdale	4A Washington	3A Cass*	2A Fayette*
4A Carroll	4A Lawrence	4A Wayne	4B Castro	3B Fisher
4A Carter	4A Lewis	4A Weakley	2A Chambers*	4B Floyd
4A Cheatham	4A Lincoln	4A White	2A Cherokee*	3B Foard
3A Chester	4A Loudon	4A Williamson	3B Childress	2A Fort Bend*
4A Claiborne	4A Macon	4A Wilson	3A Clay	3A Franklin*
4A Clay	3A Madison		4B Cochran	2A Freestone*
4A Cocke	4A Marion	**TEXAS**	3B Coke	2B Frio*
4A Coffee	4A Marshall	2A Anderson*	3B Coleman	3B Gaines
3A Crockett	4A Maury	3B Andrews	3A Collin*	2A Galveston*
4A Cumberland	4A McMinn	2A Angelina*	3B Collingsworth	3B Garza
4A Davidson	3A McNairy	2A Aransas*	2A Colorado*	3A Gillespie*
4A Decatur	4A Meigs	3A Archer	2A Comal*	3B Glasscock
4A DeKalb	4A Monroe	4B Armstrong	3A Comanche*	2A Goliad*
4A Dickson	4A Montgomery	2A Atascosa*	3B Concho	2A Gonzales*
3A Dyer	4A Moore	2A Austin*	3A Cooke	4B Gray
3A Fayette	4A Morgan	4B Bailey	2A Coryell*	3A Grayson
4A Fentress	4A Obion	2B Bandera*	3B Cottle	3A Gregg*
4A Franklin	4A Overton	2A Bastrop*	3B Crane	2A Grimes*
4A Gibson	4A Perry	3B Baylor	3B Crockett	2A Guadalupe*
4A Giles	4A Pickett	2A Bee*	3B Crosby	4B Hale
4A Grainger	4A Polk	2A Bell*	3B Culberson	3B Hall
4A Greene	4A Putnam	2A Bexar*	4B Dallam	3A Hamilton*
4A Grundy	4A Rhea	3A Blanco*	3A Dallas*	4B Hansford
4A Hamblen	4A Roane	3B Borden	3B Dawson	3B Hardeman
4A Hamilton	4A Robertson	2A Bosque*	4B Deaf Smith	2A Hardin*
4A Hancock	4A Rutherford	3A Bowie*	3A Delta	2A Harris*
3A Hardeman	4A Scott	2A Brazoria*	3A Denton*	3A Harrison*
3A Hardin	4A Sequatchie	2A Brazos*	2A DeWitt*	4B Hartley
4A Hawkins	4A Sevier	3B Brewster	3B Dickens	3B Haskell

(continued)

TABLE 301.1—continued
CLIMATE ZONES, MOISTURE REGIMES, AND WARM-HUMID DESIGNATIONS
BY STATE, COUNTY AND TERRITORY

2A Hays*	2A Liberty*	2A Polk*	2A Trinity*	5B Kane
3B Hemphill	2A Limestone*	4B Potter	2A Tyler*	5B Millard
3A Henderson*	4B Lipscomb	3B Presidio	3A Upshur*	6B Morgan
2A Hidalgo*	2A Live Oak*	3A Rains*	3B Upton	5B Piute
2A Hill*	3A Llano*	4B Randall	2B Uvalde*	6B Rich
4B Hockley	3B Loving	3B Reagan	2B Val Verde*	5B Salt Lake
3A Hood*	3B Lubbock	2B Real*	3A Van Zandt*	5B San Juan
3A Hopkins*	3B Lynn	3A Red River*	2A Victoria*	5B Sanpete
2A Houston*	2A Madison*	3B Reeves	2A Walker*	5B Sevier
3B Howard	3A Marion*	2A Refugio*	2A Waller*	6B Summit
3B Hudspeth	3B Martin	4B Roberts	3B Ward	5B Tooele
3A Hunt*	3B Mason	2A Robertson*	2A Washington*	6B Uintah
4B Hutchinson	2A Matagorda*	3A Rockwall*	2B Webb*	5B Utah
3B Irion	2B Maverick*	3B Runnels	2A Wharton*	6B Wasatch
3A Jack	3B McCulloch	3A Rusk*	3B Wheeler	3B Washington
2A Jackson*	2A McLennan*	3A Sabine*	3A Wichita	5B Wayne
2A Jasper*	2A McMullen*	3A San Augustine*	3B Wilbarger	5B Weber
3B Jeff Davis	2B Medina*	2A San Jacinto*	2A Willacy*	**VERMONT**
2A Jefferson*	3B Menard	2A San Patricio*	2A Williamson*	6A (all)
2A Jim Hogg*	3B Midland	3A San Saba*	2A Wilson*	**VIRGINIA**
2A Jim Wells*	2A Milam*	3B Schleicher	3B Winkler	4A (all)
3A Johnson*	3A Mills*	3B Scurry	3A Wise	
3B Jones	3B Mitchell	3B Shackelford	3A Wood*	**WASHINGTON**
2A Karnes*	3A Montague	3A Shelby*	4B Yoakum	5B Adams
3A Kaufman*	2A Montgomery*	4B Sherman	3A Young	5B Asotin
3A Kendall*	4B Moore	3A Smith*	2B Zapata*	5B Benton
2A Kenedy*	3A Morris*	3A Somervell*	2B Zavala*	5B Chelan
3B Kent	3B Motley	2A Starr*	**UTAH**	4C Clallam
3B Kerr	3A Nacogdoches*	3A Stephens	5B Beaver	4C Clark
3B Kimble	3A Navarro*	3B Sterling	6B Box Elder	5B Columbia
3B King	2A Newton*	3B Stonewall	6B Cache	4C Cowlitz
2B Kinney*	3B Nolan	3B Sutton	6B Carbon	5B Douglas
2A Kleberg*	2A Nueces*	4B Swisher	6B Daggett	6B Ferry
3B Knox	4B Ochiltree	3A Tarrant*	5B Davis	5B Franklin
3A Lamar*	4B Oldham	3B Taylor	6B Duchesne	5B Garfield
4B Lamb	2A Orange*	3B Terrell	5B Emery	5B Grant
3A Lampasas*	3A Palo Pinto*	3B Terry	5B Garfield	4C Grays Harbor
2B La Salle*	3A Panola*	3B Throckmorton	5B Grand	4C Island
2A Lavaca*	3A Parker*	3A Titus*	5B Iron	4C Jefferson
2A Lee*	4B Parmer	3B Tom Green	5B Juab	4C King
2A Leon*	3B Pecos	2A Travis*		4C Kitsap

(continued)

TABLE 301.1—continued
CLIMATE ZONES, MOISTURE REGIMES, AND WARM-HUMID DESIGNATIONS
BY STATE, COUNTY AND TERRITORY

5B Kittitas	4A Kanawha	6A Brown	6A Pepin	6B Park
5B Klickitat	5A Lewis	6A Buffalo	6A Pierce	5B Platte
4C Lewis	4A Lincoln	7 Burnett	6A Polk	6B Sheridan
5B Lincoln	4A Logan	6A Calumet	6A Portage	7 Sublette
4C Mason	5A Marion	6A Chippewa	7 Price	6B Sweetwater
6B Okanogan	5A Marshall	6A Clark	6A Racine	7 Teton
4C Pacific	4A Mason	6A Columbia	6A Richland	6B Uinta
6B Pend Oreille	4A McDowell	6A Crawford	6A Rock	6B Washakie
4C Pierce	4A Mercer	6A Dane	6A Rusk	6B Weston
4C San Juan	5A Mineral	6A Dodge	6A Sauk	
4C Skagit	4A Mingo	6A Door	7 Sawyer	**US**
5B Skamania	5A Monongalia	7 Douglas	6A Shawano	**TERRITORIES**
4C Snohomish	4A Monroe	6A Dunn	6A Sheboygan	
5B Spokane	4A Morgan	6A Eau Claire	6A St. Croix	**AMERICAN**
6B Stevens	5A Nicholas	7 Florence	7 Taylor	**SAMOA**
4C Thurston	5A Ohio	6A Fond du Lac	6A Trempealeau	1A (all)*
4C Wahkiakum	5A Pendleton	7 Forest	6A Vernon	
5B Walla Walla	4A Pleasants	6A Grant	7 Vilas	**GUAM**
4C Whatcom	5A Pocahontas	6A Green	6A Walworth	1A (all)*
5B Whitman	5A Preston	6A Green Lake	7 Washburn	
5B Yakima	4A Putnam	6A Iowa	6A Washington	**NORTHERN**
	5A Raleigh	7 Iron	6A Waukesha	**MARIANA**
WEST VIRGINIA	5A Randolph	6A Jackson	6A Waupaca	**ISLANDS**
5A Barbour	4A Ritchie	6A Jefferson	6A Waushara	1A (all)*
4A Berkeley	4A Roane	6A Juneau	6A Winnebago	
4A Boone	5A Summers	6A Kenosha	6A Wood	**PUERTO RICO**
4A Braxton	5A Taylor	6A Kewaunee		1A (all)*
5A Brooke	5A Tucker	6A La Crosse	**WYOMING**	
4A Cabell	4A Tyler	6A Lafayette	6B Albany	**VIRGIN ISLANDS**
4A Calhoun	5A Upshur	7 Langlade	6B Big Horn	1A (all)*
4A Clay	4A Wayne	7 Lincoln	6B Campbell	
5A Doddridge	5A Webster	6A Manitowoc	6B Carbon	
5A Fayette	5A Wetzel	6A Marathon	6B Converse	
4A Gilmer	4A Wirt	6A Marinette	6B Crook	
5A Grant	4A Wood	6A Marquette	6B Fremont	
5A Greenbrier	4A Wyoming	6A Menominee	5B Goshen	
5A Hampshire		6A Milwaukee	6B Hot Springs	
5A Hancock	**WISCONSIN**	6A Monroe	6B Johnson	
5A Hardy	6A Adams	6A Oconto	6B Laramie	
5A Harrison	7 Ashland	7 Oneida	7 Lincoln	
4A Jackson	6A Barron	6A Outagamie	6B Natrona	
4A Jefferson	7 Bayfield	6A Ozaukee	6B Niobrara	

TABLE 301.3(1)
INTERNATIONAL CLIMATE ZONE DEFINITIONS

MAJOR CLIMATE TYPE DEFINITIONS
Marine (C) Definition—Locations meeting all four criteria: 1. Mean temperature of coldest month between -3°C (27°F) and 18°C (65°F) 2. Warmest month mean < 22°C (72°F) 3. At least four months with mean temperatures over 10°C (50°F) 4. Dry season in summer. The month with the heaviest precipitation in the cold season has at least three times as much precipitation as the month with the least precipitation in the rest of the year. The cold season is October through March in the Northern Hemisphere and April through September in the Southern Hemisphere.
Dry (B) Definition—Locations meeting the following criteria: Not marine and $\quad P_{in} < 0.44 \times (TF - 19.5) \qquad [P_{cm} < 2.0 \times (TC + 7)$ in SI units] where: $\quad P_{in}$ = Annual precipitation in inches (cm) $\quad T$ = Annual mean temperature in °F (°C)
Moist (A) Definition—Locations that are not marine and not dry.
Warm-humid Definition—Moist (A) locations where either of the following wet-bulb temperature conditions shall occur during the warmest six consecutive months of the year: 1. 67°F (19.4°C) or higher for 3,000 or more hours; or 2. 73°F (22.8°C) or higher for 1,500 or more hours

For SI: °C = [(°F)-32]/1.8; 1 inch = 2.54 cm.

TABLE 301.3(2)
INTERNATIONAL CLIMATE ZONE DEFINITIONS

ZONE NUMBER	THERMAL CRITERIA	
	IP Units	SI Units
1	9000 < CDD50°F	5000 < CDD10°C
2	6300 < CDD50°F ≤ 9000	3500 < CDD10°C ≤ 5000
3A and 3B	4500 < CDD50°F ≤ 6300 AND HDD65°F ≤ 5400	2500 < CDD10°C ≤ 3500 AND HDD18°C ≤ 3000
4A and 4B	CDD50°F ≤ 4500 AND HDD65°F ≤ 5400	CDD10°C ≤ 2500 AND HDD18°C ≤ 3000
3C	HDD65°F ≤ 3600	HDD18°C ≤ 2000
4C	3600 < HDD65°F ≤ 5400	2000 < HDD18°C ≤ 3000
5	5400 < HDD65°F ≤ 7200	3000 < HDD18°C ≤ 4000
6	7200 < HDD65°F ≤ 9000	4000 < HDD18°C ≤ 5000
7	9000 < HDD65°F ≤ 12600	5000 < HDD18°C ≤ 7000
8	12600 < HDD65°F	7000 < HDD18°C

For SI: °C = [(°F)-32]/1.8

SECTION 302
DESIGN CONDITIONS

302.1 Interior design conditions. The interior design temperatures used for heating and cooling load calculations shall be a maximum of 72°F (22°C) for heating and minimum of 75°F (24°C) for cooling.

SECTION 303
MATERIALS, SYSTEMS AND EQUIPMENT

303.1 Identification. Materials, systems and equipment shall be identified in a manner that will allow a determination of compliance with the applicable provisions of this code.

303.1.1 Building thermal envelope insulation. An *R*-value identification mark shall be applied by the manufacturer to each piece of *building thermal envelope* insulation 12 inches (305 mm) or greater in width. Alternately, the insulation installers shall provide a certification listing the type, manufacturer and *R*-value of insulation installed in each element of the *building thermal envelope*. For blown or sprayed insulation (fiberglass and cellulose), the initial installed thickness, settled thickness, settled *R*-value, installed density, coverage area and number of bags installed shall be *listed* on the certification. For sprayed polyurethane foam (SPF) insulation, the installed thickness of the areas covered and *R*-value of installed thickness shall be *listed* on the certification. The insulation installer shall sign, date and post the certification in a conspicuous location on the job site.

303.1.1.1 Blown or sprayed roof/ceiling insulation. The thickness of blown-in or sprayed roof/ceiling insulation (fiberglass or cellulose) shall be written in inches (mm) on markers that are installed at least one for every 300 square feet (28 m²) throughout the attic space. The markers shall be affixed to the trusses or joists and marked with the minimum initial installed thickness with numbers a minimum of 1 inch (25 mm) in height. Each marker shall face the attic access opening. Spray polyurethane foam thickness and installed *R*-value shall be *listed* on certification provided by the insulation installer.

303.1.2 Insulation mark installation. Insulating materials shall be installed such that the manufacturer's *R*-value mark is readily observable upon inspection.

303.1.3 Fenestration product rating. *U*-factors of fenestration products (windows, doors and skylights) shall be determined in accordance with NFRC 100 by an accredited, independent laboratory, and labeled and certified by the manufacturer. Products lacking such a labeled *U*-factor shall be assigned a default *U*-factor from Table 303.1.3(1) or 303.1.3(2). The solar heat gain coefficient (SHGC) of glazed fenestration products (windows, glazed doors and skylights) shall be determined in accordance with NFRC 200 by an accredited, independent laboratory, and labeled and certified by the manufacturer. Products lacking such a labeled SHGC shall be assigned a default SHGC from Table 303.1.3(3).

TABLE 303.1.3(1)
DEFAULT GLAZED FENESTRATION *U*-FACTOR

FRAME TYPE	SINGLE PANE	DOUBLE PANE	SKYLIGHT	
			Single	Double
Metal	1.20	0.80	2.00	1.30
Metal with Thermal Break	1.10	0.65	1.90	1.10
Nonmetal or Metal Clad	0.95	0.55	1.75	1.05
Glazed Block	0.60			

TABLE 303.1.3(2)
DEFAULT DOOR *U*-FACTORS

DOOR TYPE	*U*-FACTOR
Uninsulated Metal	1.20
Insulated Metal	0.60
Wood	0.50
Insulated, nonmetal edge, max 45% glazing, any glazing double pane	0.35

TABLE 303.1.3(3)
DEFAULT GLAZED FENESTRATION SHGC

SINGLE GLAZED		DOUBLE GLAZED		
Clear	Tinted	Clear	Tinted	GLAZED BLOCK
0.8	0.7	0.7	0.6	0.6

303.1.4 Insulation product rating. The thermal resistance (*R*-value) of insulation shall be determined in accordance with the U.S. Federal Trade Commission *R*-value rule (CFR Title 16, Part 460, May 31, 2005) in units of h × ft² × °F/Btu at a mean temperature of 75°F (24°C).

303.2 Installation. All materials, systems and equipment shall be installed in accordance with the manufacturer's installation instructions and the *International Building Code*.

303.2.1 Protection of exposed foundation insulation. Insulation applied to the exterior of basement walls, crawlspace walls and the perimeter of slab-on-grade floors shall have a rigid, opaque and weather-resistant protective covering to prevent the degradation of the insulation's thermal performance. The protective covering shall cover the exposed exterior insulation and extend a minimum of 6 inches (153 mm) below grade.

303.3 Maintenance information. Maintenance instructions shall be furnished for equipment and systems that require preventive maintenance. Required regular maintenance actions shall be clearly stated and incorporated on a readily accessible label. The label shall include the title or publication number for the operation and maintenance manual for that particular model and type of product.

CHAPTER 4

RESIDENTIAL ENERGY EFFICIENCY

SECTION 401
GENERAL

401.1 Scope. This chapter applies to residential buildings.

401.2 Compliance. Projects shall comply with Sections 401, 402.4, 402.5, and 403.1, 403.2.2, 403.2.3, and 403.3 through 403.9 (referred to as the mandatory provisions) and either:

1. Sections 402.1 through 402.3, 403.2.1 and 404.1 (prescriptive); or

2. Section 405 (performance).

401.3 Certificate. A permanent certificate shall be posted on or in the electrical distribution panel. The certificate shall not cover or obstruct the visibility of the circuit directory label, service disconnect label or other required labels. The certificate shall be completed by the builder or registered design professional. The certificate shall list the predominant *R*-values of insulation installed in or on ceiling/roof, walls, foundation (slab, *basement wall*, crawlspace wall and/or floor) and ducts outside conditioned spaces; *U*-factors for fenestration and the solar heat gain coefficient (SHGC) of fenestration. Where there is more than one value for each component, the certificate shall list the value covering the largest area. The certificate shall list the types and efficiencies of heating, cooling and service water heating equipment. Where a gas-fired unvented room heater, electric furnace, or baseboard electric heater is installed in the residence, the certificate shall list "gas-fired unvented room heater," "electric furnace" or "baseboard electric heater," as appropriate. An efficiency shall not be *listed* for gas-fired unvented room heaters, electric furnaces or electric baseboard heaters.

SECTION 402
BUILDING THERMAL ENVELOPE

402.1 General (Prescriptive).

402.1.1 Insulation and fenestration criteria. The *building thermal envelope* shall meet the requirements of Table 402.1.1 based on the climate *zone* specified in Chapter 3.

402.1.2 *R*-value computation. Insulation material used in layers, such as framing cavity insulation and insulating sheathing, shall be summed to compute the component *R*-value. The manufacturer's settled *R*-value shall be used for blown insulation. Computed *R*-values shall not include an *R*-value for other building materials or air films.

TABLE 402.1.1
INSULATION AND FENESTRATION REQUIREMENTS BY COMPONENT[a]

CLIMATE ZONE	FENESTRATION *U*-FACTOR[b]	SKYLIGHT[b] *U*-FACTOR	GLAZED FENESTRATION SHGC[b, e]	CEILING *R*-VALUE	WOOD FRAME WALL *R*-VALUE	MASS WALL *R*-VALUE[i]	FLOOR *R*-VALUE	BASEMENT[c] WALL *R*-VALUE	SLAB[d] *R*-VALUE & DEPTH	CRAWL SPACE[c] WALL *R*-VALUE
1	1.2	0.75	0.30	30	13	3/4	13	0	0	0
2	0.65[j]	0.75	0.30	30	13	4/6	13	0	0	0
3	0.50[j]	0.65	0.30	30	13	5/8	19	5/13[f]	0	5/13
4 except Marine	0.35	0.60	NR	38	13	5/10	19	10/13	10, 2 ft	10/13
5 and Marine 4	0.35	0.60	NR	38	20 or 13+5[h]	13/17	30[g]	10/13	10, 2 ft	10/13
6	0.35	0.60	NR	49	20 or 13+5[h]	15/19	30[g]	15/19	10, 4 ft	10/13
7 and 8	0.35	0.60	NR	49	21	19/21	38[g]	15/19	10, 4 ft	10/13

For SI: 1 foot = 304.8 mm.

a. *R*-values are minimums. *U*-factors and SHGC are maximums. R-19 batts compressed into a nominal 2 × 6 framing cavity such that the *R*-value is reduced by R-1 or more shall be marked with the compressed batt *R*-value in addition to the full thickness *R*-value.

b. The fenestration *U*-factor column excludes skylights. The SHGC column applies to all glazed fenestration.

c. "15/19" means R-15 continuous insulated sheathing on the interior or exterior of the home or R-19 cavity insulation at the interior of the basement wall. "15/19" shall be permitted to be met with R-13 cavity insulation on the interior of the basement wall plus R-5 continuous insulated sheathing on the interior or exterior of the home. "10/13" means R-10 continuous insulated sheathing on the interior or exterior of the home or R-13 cavity insulation at the interior of the basement wall.

d. R-5 shall be added to the required slab edge *R*-values for heated slabs. Insulation depth shall be the depth of the footing or 2 feet, whichever is less in Zones 1 through 3 for heated slabs.

e. There are no SHGC requirements in the Marine Zone.

f. Basement wall insulation is not required in warm-humid locations as defined by Figure 301.1 and Table 301.1.

g. Or insulation sufficient to fill the framing cavity, R-19 minimum.

h. "13+5" means R-13 cavity insulation plus R-5 insulated sheathing. If structural sheathing covers 25 percent or less of the exterior, insulating sheathing is not required where structural sheathing is used. If structural sheathing covers more than 25 percent of exterior, structural sheathing shall be supplemented with insulated sheathing of at least R-2.

i. The second *R*-value applies when more than half the insulation is on the interior of the mass wall.

j. For impact rated fenestration complying with Section R301.2.1.2 of the *International Residential Code* or Section 1609.1.2 of the *International Building Code*, the maximum *U*-factor shall be 0.75 in Zone 2 and 0.65 in Zone 3.

TABLE 402.1.3
EQUIVALENT U-FACTORS[a]

CLIMATE ZONE	FENESTRATION U-FACTOR	SKYLIGHT U-FACTOR	CEILING U-FACTOR	FRAME WALL U-FACTOR	MASS WALL U-FACTOR[b]	FLOOR U-FACTOR	BASEMENT WALL U-FACTOR	CRAWL SPACE WALL U-FACTOR[c]
1	1.20	0.75	0.035	0.082	0.197	0.064	0.360	0.477
2	0.65	0.75	0.035	0.082	0.165	0.064	0.360	0.477
3	0.50	0.65	0.035	0.082	0.141	0.047	0.091[c]	0.136
4 except Marine	0.35	0.60	0.030	0.082	0.141	0.047	0.059	0.065
5 and Marine 4	0.35	0.60	0.030	0.057	0.082	0.033	0.059	0.065
6	0.35	0.60	0.026	0.057	0.060	0.033	0.050	0.065
7 and 8	0.35	0.60	0.026	0.057	0.057	0.028	0.050	0.065

a. Nonfenestration U-factors shall be obtained from measurement, calculation or an approved source.
b. When more than half the insulation is on the interior, the mass wall U-factors shall be a maximum of 0.17 in Zone 1, 0.14 in Zone 2, 0.12 in Zone 3, 0.10 in Zone 4 except Marine, and the same as the frame wall U-factor in Marine Zone 4 and Zones 5 through 8.
c. Basement wall U-factor of 0.360 in warm-humid locations as defined by Figure 301.1 and Table 301.1.

402.1.3 U-factor alternative. An assembly with a U-factor equal to or less than that specified in Table 402.1.3 shall be permitted as an alternative to the R-value in Table 402.1.1.

402.1.4 Total UA alternative. If the total *building thermal envelope* UA (sum of U-factor times assembly area) is less than or equal to the total UA resulting from using the U-factors in Table 402.1.3 (multiplied by the same assembly area as in the proposed building), the building shall be considered in compliance with Table 402.1.1. The UA calculation shall be done using a method consistent with the ASHRAE *Handbook of Fundamentals* and shall include the thermal bridging effects of framing materials. The SHGC requirements shall be met in addition to UA compliance.

402.2 Specific insulation requirements (Prescriptive).

402.2.1 Ceilings with attic spaces. When Section 402.1.1 would require R-38 in the ceiling, R-30 shall be deemed to satisfy the requirement for R-38 wherever the full height of uncompressed R-30 insulation extends over the wall top plate at the eaves. Similarly, R-38 shall be deemed to satisfy the requirement for R-49 wherever the full height of uncompressed R-38 insulation extends over the wall top plate at the eaves. This reduction shall not apply to the U-factor alternative approach in Section 402.1.3 and the total UA alternative in Section 402.1.4.

402.2.2 Ceilings without attic spaces. Where Section 402.1.1 would require insulation levels above R-30 and the design of the roof/ceiling assembly does not allow sufficient space for the required insulation, the minimum required insulation for such roof/ceiling assemblies shall be R-30. This reduction of insulation from the requirements of Sec-tion 402.1.1 shall be limited to 500 square feet (46 m²) or 20 percent of the total insulated ceiling area, whichever is less. This reduction shall not apply to the U-factor alternative approach in Section 402.1.3 and the total UA alternative in Section 402.1.4.

402.2.3 Access hatches and doors. Access doors from conditioned spaces to unconditioned spaces (e.g., attics and crawl spaces) shall be weatherstripped and insulated to a level equivalent to the insulation on the surrounding surfaces. Access shall be provided to all equipment that prevents damaging or compressing the insulation. A wood framed or equivalent baffle or retainer is required to be provided when loose fill insulation is installed, the purpose of which is to prevent the loose fill insulation from spilling into the living space when the attic access is opened, and to provide a permanent means of maintaining the installed R-value of the loose fill insulation.

402.2.4 Mass walls. Mass walls for the purposes of this chapter shall be considered above-grade walls of concrete block, concrete, insulated concrete form (ICF), masonry cavity, brick (other than brick veneer), earth (adobe, compressed earth block, rammed earth) and solid timber/logs.

402.2.5 Steel-frame ceilings, walls, and floors. Steel-frame ceilings, walls and floors shall meet the insulation requirements of Table 402.2.5 or shall meet the U-factor requirements in Table 402.1.3. The calculation of the U-factor for a steel-frame envelope assembly shall use a series-parallel path calculation method.

Exception: In Climate Zones 1 and 2, the continuous insulation requirements in Table 402.2.5 shall be permitted to be reduced to R-3 for steel frame wall assemblies with studs spaced at 24 inches (610 mm) on center.

TABLE 402.2.5
STEEL-FRAME CEILING, WALL AND FLOOR INSULATION
(R-VALUE)

WOOD FRAME R-VALUE REQUIREMENT	COLD-FORMED STEEL EQUIVALENT R-VALUE[a]
Steel Truss Ceilings[b]	
R-30	R-38 or R-30 + 3 or R-26 + 5
R-38	R-49 or R-38 + 3
R-49	R-38 + 5
Steel Joist Ceilings[b]	
R-30	R-38 in 2×4 or 2×6 or 2×8 R-49 in any framing
R-38	R-49 in 2×4 or 2×6 or 2×8 or 2×10
Steel-Framed Wall	
R-13	R-13 + 5 or R-15 + 4 or R-21 + 3 or R-0 + 10
R-19	R-13 + 9 or R-19 + 8 or R-25 + 7
R-21	R-13 + 10 or R-19 + 9 or R-25 + 8
Steel Joist Floor	
R-13	R-19 in 2×6 R-19 + 6 in 2×8 or 2×10
R-19	R-19 + 6 in 2×6 R-19 + 12 in 2×8 or 2×10

a. Cavity insulation R-value is listed first, followed by continuous insulation R-value.

b. Insulation exceeding the height of the framing shall cover the framing.

402.2.6 Floors. Floor insulation shall be installed to maintain permanent contact with the underside of the subfloor decking.

402.2.7 Basement walls. Walls associated with conditioned basements shall be insulated from the top of the *basement wall* down to 10 feet (3048 mm) below grade or to the basement floor, whichever is less. Walls associated with unconditioned basements shall meet this requirement unless the floor overhead is insulated in accordance with Sections 402.1.1 and 402.2.6.

402.2.8 Slab-on-grade floors. Slab-on-grade floors with a floor surface less than 12 inches (305 mm) below grade shall be insulated in accordance with Table 402.1.1. The insulation shall extend downward from the top of the slab on the outside or inside of the foundation wall. Insulation located below grade shall be extended the distance provided in Table 402.1.1 by any combination of vertical insulation, insulation extending under the slab or insulation extending out from the building. Insulation extending away from the building shall be protected by pavement or by a minimum of 10 inches (254 mm) of soil. The top edge of the insulation installed between the *exterior wall* and the edge of the interior slab shall be permitted to be cut at a 45-degree (0.79 rad) angle away from the *exterior wall*. Slab-edge insulation is not required in jurisdictions designated by the *code official* as having a very heavy termite infestation.

402.2.9 Crawl space walls. As an alternative to insulating floors over crawl spaces, crawl space walls shall be permitted to be insulated when the crawl space is not vented to the outside. Crawl space wall insulation shall be permanently fastened to the wall and extend downward from the floor to the finished grade level and then vertically and/or horizon-

tally for at least an additional 24 inches (610 mm). Exposed earth in unvented crawl space foundations shall be covered with a continuous Class I vapor retarder in accordance with the *International Building Code*. All joints of the vapor retarder shall overlap by 6 inches (153 mm) and be sealed or taped. The edges of the vapor retarder shall extend at least 6 inches (153 mm) up the stem wall and shall be attached to the stem wall.

402.2.10 Masonry veneer. Insulation shall not be required on the horizontal portion of the foundation that supports a masonry veneer.

402.2.11 Thermally isolated sunroom insulation. The minimum ceiling insulation R-values shall be R-19 in Zones 1 through 4 and R-24 in Zones 5 through 8. The minimum wall R-value shall be R-13 in all zones. New wall(s) separating a sunroom from *conditioned space* shall meet the *building thermal envelope* requirements.

402.3 Fenestration. (Prescriptive).

402.3.1 U-factor. An area-weighted average of fenestration products shall be permitted to satisfy the U-factor requirements.

402.3.2 Glazed fenestration SHGC. An area-weighted average of fenestration products more than 50 percent glazed shall be permitted to satisfy the SHGC requirements.

402.3.3 Glazed fenestration exemption. Up to 15 square feet (1.4 m²) of glazed fenestration per dwelling unit shall be permitted to be exempt from U-factor and SHGC requirements in Section 402.1.1. This exemption shall not apply to the U-factor alternative approach in Section 402.1.3 and the Total UA alternative in Section 402.1.4.

402.3.4 Opaque door exemption. One side-hinged opaque door assembly up to 24 square feet (2.22 m²) in area is exempted from the U-factor requirement in Section 402.1.1. This exemption shall not apply to the U-factor alternative approach in Section 402.1.3 and the total UA alternative in Section 402.1.4.

402.3.5 Thermally isolated sunroom U-factor. For Zones 4 through 8, the maximum fenestration U-factor shall be 0.50 and the maximum skylight U-factor shall be 0.75. New windows and doors separating the sunroom from *conditioned space* shall meet the *building thermal envelope* requirements.

402.3.6 Replacement fenestration. Where some or all of an existing fenestration unit is replaced with a new fenestration product, including sash and glazing, the replacement fenestration unit shall meet the applicable requirements for U-factor and SHGC in Table 402.1.1.

402.4 Air leakage (Mandatory).

402.4.1 Building thermal envelope. The *building thermal envelope* shall be durably sealed to limit infiltration. The sealing methods between dissimilar materials shall allow for differential expansion and contraction. The following shall be caulked, gasketed, weatherstripped or otherwise sealed with an air barrier material, suitable film or solid material:

1. All joints, seams and penetrations.

2. Site-built windows, doors and skylights.

3. Openings between window and door assemblies and their respective jambs and framing.

4. Utility penetrations.

5. Dropped ceilings or chases adjacent to the thermal envelope.

6. Knee walls.

7. Walls and ceilings separating a garage from conditioned spaces.

8. Behind tubs and showers on exterior walls.

9. Common walls between dwelling units.

10. Attic access openings.

11. Rim joist junction.

12. Other sources of infiltration.

402.4.2 Air sealing and insulation. Building envelope air tightness and insulation installation shall be demonstrated to comply with one of the following options given by Section 402.4.2.1 or 402.4.2.2:

402.4.2.1 Testing option. Building envelope tightness and insulation installation shall be considered acceptable when tested air leakage is less than seven air changes per hour (ACH) when tested with a blower door at a pressure of 50 pascals (1 psf). Testing shall occur after rough in and after installation of penetrations of the building envelope, including penetrations for utilities, plumbing, electrical, ventilation and combustion appliances.

During testing:

1. Exterior windows and doors, fireplace and stove doors shall be closed, but not sealed;

2. Dampers shall be closed, but not sealed, including exhaust, intake, makeup air, backdraft and flue dampers;

3. Interior doors shall be open;

4. Exterior openings for continuous ventilation systems and heat recovery ventilators shall be closed and sealed;

5. Heating and cooling system(s) shall be turned off;

6. HVAC ducts shall not be sealed; and

7. Supply and return registers shall not be sealed.

402.4.2.2 Visual inspection option. Building envelope tightness and insulation installation shall be considered acceptable when the items listed in Table 402.4.2, applicable to the method of construction, are field verified. Where required by the *code official*, an *approved* party independent from the installer of the insulation shall inspect the air barrier and insulation.

402.4.3 Fireplaces. New wood-burning fireplaces shall have gasketed doors and outdoor combustion air.

402.4.4 Fenestration air leakage. Windows, skylights and sliding glass doors shall have an air infiltration rate of no more than 0.3 cfm per square foot (1.5 L/s/m²), and swinging doors no more than 0.5 cfm per square foot (2.6 L/s/m²), when tested according to NFRC 400 or AAMA/WDMA/CSA 101/I.S.2/A440 by an accredited, independent laboratory and *listed* and *labeled* by the manufacturer.

Exceptions: Site-built windows, skylights and doors.

402.4.5 Recessed lighting. Recessed luminaires installed in the *building thermal envelope* shall be sealed to limit air leakage between conditioned and unconditioned spaces. All recessed luminaires shall be IC-rated and *labeled* as meeting ASTM E 283 when tested at 1.57 psf (75 Pa) pressure differential with no more than 2.0 cfm (0.944 L/s) of air movement from the *conditioned space* to the ceiling cavity. All recessed luminaires shall be sealed with a gasket or caulk between the housing and the interior wall or ceiling covering.

402.5 Maximum fenestration *U*-factor and SHGC (Mandatory). The area-weighted average maximum fenestration *U*-factor permitted using trade-offs from Section 402.1.4 or 405 shall be 0.48 in Zones 4 and 5 and 0.40 in Zones 6 through 8 for vertical fenestration, and 0.75 in Zones 4 through 8 for skylights. The area-weighted average maximum fenestration SHGC permitted using trade-offs from Section 405 in Zones 1 through 3 shall be 0.50.

SECTION 403
SYSTEMS

403.1 Controls (Mandatory). At least one thermostat shall be provided for each separate heating and cooling system.

403.1.1 Programmable thermostat. Where the primary heating system is a forced-air furnace, at least one thermostat per dwelling unit shall be capable of controlling the heating and cooling system on a daily schedule to maintain different temperature set points at different times of the day. This thermostat shall include the capability to set back or temporarily operate the system to maintain zone temperatures down to 55°F (13°C) or up to 85°F (29°C). The thermostat shall initially be programmed with a heating temperature set point no higher than 70°F (21°C) and a cooling temperature set point no lower than 78°F (26°C).

403.1.2 Heat pump supplementary heat (Mandatory). Heat pumps having supplementary electric-resistance heat shall have controls that, except during defrost, prevent supplemental heat operation when the heat pump compressor can meet the heating load.

403.2 Ducts.

403.2.1 Insulation (Prescriptive). Supply ducts in attics shall be insulated to a minimum of R-8. All other ducts shall be insulated to a minimum of R-6.

Exception: Ducts or portions thereof located completely inside the *building thermal envelope*.

403.2.2 Sealing (Mandatory). All ducts, air handlers, filter boxes and building cavities used as ducts shall be sealed.

Joints and seams shall comply with Section M1601.4.1 of the *International Residential Code*.

Duct tightness shall be verified by either of the following:

1. Postconstruction test: Leakage to outdoors shall be less than or equal to 8 cfm (226.5 L/min) per 100 ft² (9.29 m²) of *conditioned floor area* or a total leakage less than or equal to 12 cfm (12 L/min) per 100 ft² (9.29 m²) of *conditioned floor area* when tested at a pressure differential of 0.1 inches w.g. (25 Pa) across the entire system, including the manufacturer's air handler enclosure. All register boots shall be taped or otherwise sealed during the test.

2. Rough-in test: Total leakage shall be less than or equal to 6 cfm (169.9 L/min) per 100 ft² (9.29 m²) of *conditioned floor area* when tested at a pressure differential of 0.1 inches w.g. (25 Pa) across the roughed in system, including the manufacturer's air handler enclosure. All register boots shall be taped or otherwise sealed during the test. If the air handler is not installed at the time of the test, total leakage shall be less than or equal to 4 cfm (113.3 L/min) per 100 ft² (9.29 m²) of *conditioned floor area*.

Exceptions: Duct tightness test is not required if the air handler and all ducts are located within *conditioned space*.

TABLE 402.4.2
AIR BARRIER AND INSULATION INSPECTION COMPONENT CRITERIA

COMPONENT	CRITERIA
Air barrier and thermal barrier	Exterior thermal envelope insulation for framed walls is installed in substantial contact and continuous alignment with building envelope air barrier. Breaks or joints in the air barrier are filled or repaired. Air-permeable insulation is not used as a sealing material. Air-permeable insulation is inside of an air barrier.
Ceiling/attic	Air barrier in any dropped ceiling/soffit is substantially aligned with insulation and any gaps are sealed. Attic access (except unvented attic), knee wall door, or drop down stair is sealed.
Walls	Corners and headers are insulated. Junction of foundation and sill plate is sealed.
Windows and doors	Space between window/door jambs and framing is sealed.
Rim joists	Rim joists are insulated and include an air barrier.
Floors (including above-garage and cantilevered floors)	Insulation is installed to maintain permanent contact with underside of subfloor decking. Air barrier is installed at any exposed edge of insulation.
Crawl space walls	Insulation is permanently attached to walls. Exposed earth in unvented crawl spaces is covered with Class I vapor retarder with overlapping joints taped.
Shafts, penetrations	Duct shafts, utility penetrations, knee walls and flue shafts opening to exterior or unconditioned space are sealed.
Narrow cavities	Batts in narrow cavities are cut to fit, or narrow cavities are filled by sprayed/blown insulation.
Garage separation	Air sealing is provided between the garage and conditioned spaces.
Recessed lighting	Recessed light fixtures are air tight, IC rated, and sealed to drywall. Exception—fixtures in conditioned space.
Plumbing and wiring	Insulation is placed between outside and pipes. Batt insulation is cut to fit around wiring and plumbing, or sprayed/blown insulation extends behind piping and wiring.
Shower/tub on exterior wall	Showers and tubs on exterior walls have insulation and an air barrier separating them from the exterior wall.
Electrical/phone box on exterior walls	Air barrier extends behind boxes or air sealed-type boxes are installed.
Common wall	Air barrier is installed in common wall between dwelling units.
HVAC register boots	HVAC register boots that penetrate building envelope are sealed to subfloor or drywall.
Fireplace	Fireplace walls include an air barrier.

403.2.3 Building cavities (Mandatory). Building framing cavities shall not be used as supply ducts.

403.3 Mechanical system piping insulation (Mandatory). Mechanical system piping capable of carrying fluids above 105°F (41°C) or below 55°F (13°C) shall be insulated to a minimum of R-3.

403.4 Circulating hot water systems (Mandatory). All circulating service hot water piping shall be insulated to at least R-2. Circulating hot water systems shall include an automatic or readily *accessible* manual switch that can turn off the hot-water circulating pump when the system is not in use.

403.5 Mechanical ventilation (Mandatory). Outdoor air intakes and exhausts shall have automatic or gravity dampers that close when the ventilation system is not operating.

403.6 Equipment sizing (Mandatory). Heating and cooling equipment shall be sized in accordance with Section M1401.3 of the *International Residential Code*.

403.7 Systems serving multiple dwelling units (Mandatory). Systems serving multiple dwelling units shall comply with Sections 503 and 504 in lieu of Section 403.

403.8 Snow melt system controls (Mandatory). Snow- and ice-melting systems, supplied through energy service to the building, shall include automatic controls capable of shutting off the system when the pavement temperature is above 50°F, and no precipitation is falling and an automatic or manual control that will allow shutoff when the outdoor temperature is above 40°F.

403.9 Pools (Mandatory). Pools shall be provided with energy-conserving measures in accordance with Sections 403.9.1 through 403.9.3.

403.9.1 Pool heaters. All pool heaters shall be equipped with a readily *accessible* on-off switch to allow shutting off the heater without adjusting the thermostat setting. Pool heaters fired by natural gas or LPG shall not have continuously burning pilot lights.

403.9.2 Time switches. Time switches that can automatically turn off and on heaters and pumps according to a preset schedule shall be installed on swimming pool heaters and pumps.

Exceptions:

1. Where public health standards require 24-hour pump operation.

2. Where pumps are required to operate solar- and waste-heat-recovery pool heating systems.

403.9.3 Pool covers. Heated pools shall be equipped with a vapor-retardant pool cover on or at the water surface. Pools heated to more than 90°F (32°C) shall have a pool cover with a minimum insulation value of R-12.

Exception: Pools deriving over 60 percent of the energy for heating from site-recovered energy or solar energy source.

SECTION 404
ELECTRICAL POWER AND LIGHTING SYSTEMS

404.1 Lighting equipment. A minimum of 50 percent of the lamps in permanently installed lighting fixtures shall be high-efficacy lamps.

SECTION 405
SIMULATED PERFORMANCE ALTERNATIVE
(Performance)

405.1 Scope. This section establishes criteria for compliance using simulated energy performance analysis. Such analysis shall include heating, cooling, and service water heating energy only.

405.2 Mandatory requirements. Compliance with this section requires that the mandatory provisions identified in Section 401.2 be met. All supply and return ducts not completely inside the *building thermal envelope* shall be insulated to a minimum of R-6.

405.3 Performance-based compliance. Compliance based on simulated energy performance requires that a proposed residence (*proposed design*) be shown to have an annual energy cost that is less than or equal to the annual energy cost of the *standard reference design*. Energy prices shall be taken from a source *approved* by the *code official*, such as the Department of Energy, Energy Information Administration's *State Energy Price and Expenditure Report*. *Code officials* shall be permitted to require time-of-use pricing in energy cost calculations.

Exception: The energy use based on source energy expressed in Btu or Btu per square foot of *conditioned floor area* shall be permitted to be substituted for the energy cost. The source energy multiplier for electricity shall be 3.16. The source energy multiplier for fuels other than electricity shall be 1.1.

405.4 Documentation.

405.4.1 Compliance software tools. Documentation verifying that the methods and accuracy of the compliance software tools conform to the provisions of this section shall be provided to the *code official*.

405.4.2 Compliance report. Compliance software tools shall generate a report that documents that the *proposed design* complies with Section 405.3. The compliance documentation shall include the following information:

1. Address or other identification of the residence;

2. An inspection checklist documenting the building component characteristics of the *proposed design* as listed in Table 405.5.2(1). The inspection checklist shall show results for both the *standard reference design* and the *proposed design*, and shall document all inputs entered by the user necessary to reproduce the results;

3. Name of individual completing the compliance report; and

4. Name and version of the compliance software tool.

> **Exception:** Multiple orientations. When an otherwise identical building model is offered in multiple orientations, compliance for any orientation shall be permitted by documenting that the building meets the performance requirements in each of the four cardinal (north, east, south and west) orientations.

405.4.3 Additional documentation. The *code official* shall be permitted to require the following documents:

1. Documentation of the building component characteristics of the *standard reference design*.

2. A certification signed by the builder providing the building component characteristics of the *proposed design* as given in Table 405.5.2(1).

3. Documentation of the actual values used in the software calculations for the *proposed design*.

405.5 Calculation procedure.

405.5.1 General. Except as specified by this section, the *standard reference design* and *proposed design* shall be configured and analyzed using identical methods and techniques.

405.5.2 Residence specifications. The *standard reference design* and *proposed design* shall be configured and analyzed as specified by Table 405.5.2(1). Table 405.5.2(1) shall include by reference all notes contained in Table 402.1.1.

405.6 Calculation software tools.

405.6.1 Minimum capabilities. Calculation procedures used to comply with this section shall be software tools capable of calculating the annual energy consumption of all building elements that differ between the *standard reference design* and the *proposed design* and shall include the following capabilities:

1. Computer generation of the *standard reference design* using only the input for the *proposed design*. The calculation procedure shall not allow the user to directly modify the building component characteristics of the *standard reference design*.

2. Calculation of whole-building (as a single *zone*) sizing for the heating and cooling equipment in the *standard reference design* residence in accordance with Section M1401.3 of the *International Residential Code*.

3. Calculations that account for the effects of indoor and outdoor temperatures and part-load ratios on the performance of heating, ventilating and air-conditioning equipment based on climate and equipment sizing.

4. Printed *code official* inspection checklist listing each of the *proposed design* component characteristics from Table 405.5.2(1) determined by the analysis to provide compliance, along with their respective performance ratings (e.g., *R*-value, *U*-factor, SHGC, HSPF, AFUE, SEER, EF, etc.).

405.6.2 Specific approval. Performance analysis tools meeting the applicable sections of Section 405 shall be permitted to be *approved*. Tools are permitted to be *approved* based on meeting a specified threshold for a jurisdiction. The *code official* shall be permitted to approve tools for a specified application or limited scope.

405.6.3 Input values. When calculations require input values not specified by Sections 402, 403, 404 and 405, those input values shall be taken from an *approved* source.

TABLE 405.5.2(1)
SPECIFICATIONS FOR THE STANDARD REFERENCE AND PROPOSED DESIGNS

BUILDING COMPONENT	STANDARD REFERENCE DESIGN	PROPOSED DESIGN
Above-grade walls	Type: mass wall if proposed wall is mass; otherwise wood frame. Gross area: same as proposed U-factor: from Table 402.1.3 Solar absorptance = 0.75 Emittance = 0.90	As proposed As proposed As proposed As proposed As proposed
Basement and crawl space walls	Type: same as proposed Gross area: same as proposed U-factor: from Table 402.1.3, with insulation layer on interior side of walls.	As proposed As proposed As proposed
Above-grade floors	Type: wood frame Gross area: same as proposed U-factor: from Table 402.1.3	As proposed As proposed As proposed
Ceilings	Type: wood frame Gross area: same as proposed U-factor: from Table 402.1.3	As proposed As proposed As proposed
Roofs	Type: composition shingle on wood sheathing Gross area: same as proposed Solar absorptance = 0.75 Emittance = 0.90	As proposed As proposed As proposed As proposed
Attics	Type: vented with aperture = 1 ft^2 per 300 ft^2 ceiling area	As proposed
Foundations	Type: same as proposed foundation wall area above and below grade and soil characteristics: same as proposed.	As proposed As proposed
Doors	Area: 40 ft^2 Orientation: North U-factor: same as fenestration from Table 402.1.3.	As proposed As proposed As proposed
Glazing[a]	Total area[b] = (a) The proposed glazing area; where proposed glazing area is less than 15% of the conditioned floor area. (b) 15% of the conditioned floor area; where the proposed glazing area is 15% or more of the conditioned floor area. Orientation: equally distributed to four cardinal compass orientations (N, E, S & W). U-factor: from Table 402.1.3 SHGC: From Table 402.1.1 except that for climates with no requirement (NR) SHGC = 0.40 shall be used. Interior shade fraction: Summer (all hours when cooling is required) = 0.70 Winter (all hours when heating is required) = 0.85[c] External shading: none	As proposed As proposed As proposed Same as standard reference design As proposed
Skylights	None	As proposed
Thermally isolated sunrooms	None	As proposed

(continued)

TABLE 405.5.2(1)—continued
SPECIFICATIONS FOR THE STANDARD REFERENCE AND PROPOSED DESIGNS

BUILDING COMPONENT	STANDARD REFERENCE DESIGN	PROPOSED DESIGN
Air exchange rate	Specific leakage area (SLA)[e] = 0.00036 assuming no energy recovery	For residences that are not tested, the same as the standard reference design. For residences without mechanical ventilation that are tested in accordance with ASHRAE 119, Section 5.1, the measured air exchange rate[f] but not less than 0.35 ACH For residences with mechanical ventilation that are tested in accordance with ASHRAE 119, Section 5.1, the measured air exchange rate[e] combined with the mechanical ventilation rate, f which shall not be less than $0.01 \times CFA + 7.5 \times (N_{br}+1)$ where: CFA = conditioned floor area N_{br} = number of bedrooms
Mechanical ventilation	None, except where mechanical ventilation is specified by the proposed design, in which case: Annual vent fan energy use: kWh/yr = $0.03942 \times CFA + 29.565 \times (N_{br}+1)$ where: CFA = conditioned floor area N_{br} = number of bedrooms	As proposed
Internal gains	IGain = $17,900 + 23.8 \times CFA + 4104 \times N_{br}$ (Btu/day per dwelling unit)	Same as standard reference design
Internal mass	An internal mass for furniture and contents of 8 pounds per square foot of floor area.	Same as standard reference design, plus any additional mass specifically designed as a thermal storage element[g] but not integral to the building envelope or structure
Structural mass	For masonry floor slabs, 80% of floor area covered by R-2 carpet and pad, and 20% of floor directly exposed to room air. For masonry basement walls, as proposed, but with insulation required by Table 402.1.3 located on the interior side of the walls For other walls, for ceilings, floors, and interior walls, wood frame construction	As proposed As proposed As proposed
Heating systems[h]	As proposed Capacity: sized in accordance with Section M1401.3 of the *International Residential Code*	As proposed
Cooling systems[h, j]	As proposed Capacity: sized in accordance with Section M1401.3 of the *International Residential Code*	As proposed
Service H$_2$O heating[h, k, i]	As proposed Use: same as proposed design	As proposed gal/day = $30 + (10 \times N_{br})$
Thermal distribution systems	A thermal distribution system efficiency (DSE) of 0.88 shall be applied to both the heating and cooling system efficiencies for all systems other than tested duct systems. Duct insulation: From Section 403.2.1. For tested duct systems, the leakage rate shall be the applicable maximum rate from Section 403.2.2.	As tested or as specified in Table 405.5.2(2) if not tested
Thermostat	Type: Manual, cooling temperature setpoint = 75°F; Heating temperature setpoint = 72°F	Same as standard reference

(continued)

TABLE 405.5.2(1)—continued

For SI: 1 square foot = 0.93 m^2; 1 British thermal unit = 1055 J; 1 pound per square foot = 4.88 kg/m^2; 1 gallon (U.S.) = 3.785 L; °C = (°F-3)/1.8, 1 degree = 0.79 rad.

a. Glazing shall be defined as sunlight-transmitting fenestration, including the area of sash, curbing or other framing elements, that enclose conditioned space. Glazing includes the area of sunlight-transmitting fenestration assemblies in walls bounding conditioned basements. For doors where the sunlight-transmitting opening is less than 50 percent of the door area, the glazing area is the sunlight transmitting opening area. For all other doors, the glazing area is the rough frame opening area for the door including the door and the frame.

b. For residences with conditioned basements, R-2 and R-4 residences and townhouses, the following formula shall be used to determine glazing area:

$AF = A_s \times FA \times F$

where:

AF = Total glazing area.

A_s = Standard reference design total glazing area.

FA = (Above-grade thermal boundary gross wall area)/(above-grade boundary wall area + 0.5 × below-grade boundary wall area).

F = (Above-grade thermal boundary wall area)/(above-grade thermal boundary wall area + common wall area) or 0.56, whichever is greater.

and where:

Thermal boundary wall is any wall that separates conditioned space from unconditioned space or ambient conditions.

Above-grade thermal boundary wall is any thermal boundary wall component not in contact with soil.

Below-grade boundary wall is any thermal boundary wall in soil contact.

Common wall area is the area of walls shared with an adjoining dwelling unit.

c. For fenestrations facing within 15 degrees (0.26 rad) of true south that are directly coupled to thermal storage mass, the winter interior shade fraction shall be permitted to be increased to 0.95 in the proposed design.

d. Where leakage area (L) is defined in accordance with Section 5.1 of ASHRAE 119 and where:

$SLA = L/CFA$

where L and CFA are in the same units.

e. Tested envelope leakage shall be determined and documented by an independent party approved by the *code official*. Hourly calculations as specified in the 2001 ASHRAE *Handbook of Fundamentals*, Chapter 26, page 26.21, Equation 40 (Sherman-Grimsrud model) or the equivalent shall be used to determine the energy loads resulting from infiltration.

f. The combined air exchange rate for infiltration and mechanical ventilation shall be determined in accordance with Equation 43 of 2001 ASHRAE *Handbook of Fundamentals*, page 26.24 and the "Whole-house Ventilation" provisions of 2001 ASHRAE *Handbook of Fundamentals*, page 26.19 for intermittent mechanical ventilation.

g. Thermal storage element shall mean a component not part of the floors, walls or ceilings that is part of a passive solar system, and that provides thermal storage such as enclosed water columns, rock beds, or phase-change containers. A thermal storage element must be in the same room as fenestration that faces within 15 degrees (0.26 rad) of true south, or must be connected to such a room with pipes or ducts that allow the element to be actively charged.

h. For a proposed design with multiple heating, cooling or water heating systems using different fuel types, the applicable standard reference design system capacities and fuel types shall be weighted in accordance with their respective loads as calculated by accepted engineering practice for each equipment and fuel type present.

i. For a proposed design without a proposed heating system, a heating system with the prevailing federal minimum efficiency shall be assumed for both the standard reference design and proposed design. For electric heating systems, the prevailing federal minimum efficiency air-source heat pump shall be used for the standard reference design.

j. For a proposed design home without a proposed cooling system, an electric air conditioner with the prevailing federal minimum efficiency shall be assumed for both the standard reference design and the proposed design.

k. For a proposed design with a nonstorage-type water heater, a 40-gallon storage-type water heater with the prevailing federal minimum energy factor for the same fuel as the predominant heating fuel type shall be assumed. For the case of a proposed design without a proposed water heater, a 40-gallon storage-type water heater with the prevailing federal minimum efficiency for the same fuel as the predominant heating fuel type shall be assumed for both the proposed design and standard reference design.

TABLE 405.5.2(2)
DEFAULT DISTRIBUTION SYSTEM EFFICIENCIES FOR PROPOSED DESIGNS[a]

DISTRIBUTION SYSTEM CONFIGURATION AND CONDITION:	FORCED AIR SYSTEMS	HYDRONIC SYSTEMS[b]
Distribution system components located in unconditioned space	—	0.95
Untested distribution systems entirely located in conditioned space[c]	0.88	1
"Ductless" systems[d]	1	—

For SI: 1 cubic foot per minute = 0.47 L/s; 1 square foot = 0.093 m^2; 1 pound per square inch = 6895 Pa; 1 inch water gauge = 1250 Pa.

a. Default values given by this table are for untested distribution systems, which must still meet minimum requirements for duct system insulation.

b. Hydronic systems shall mean those systems that distribute heating and cooling energy directly to individual spaces using liquids pumped through closed loop piping and that do not depend on ducted, forced airflow to maintain space temperatures.

c. Entire system in conditioned space shall mean that no component of the distribution system, including the air handler unit, is located outside of the conditioned space.

d. Ductless systems shall be allowed to have forced airflow across a coil but shall not have any ducted airflow external to the manufacturer's air handler enclosure.

CHAPTER 5

COMMERCIAL ENERGY EFFICIENCY

SECTION 501
GENERAL

501.1 Scope. The requirements contained in this chapter are applicable to commercial buildings, or portions of commercial buildings. These commercial buildings shall meet either the requirements of ANSI/ASHRAE/IESNA Standard 90.1, *Energy Standard for Buildings Except for Low-Rise Residential Buildings*, or the requirements contained in this chapter.

501.2 Application. The *commercial building* project shall comply with the requirements in Sections 502 (Building envelope requirements), 503 (Building mechanical systems), 504 (Service water heating) and 505 (Electrical power and lighting systems) in its entirety. As an alternative the *commercial building* project shall comply with the requirements of ANSI/ASHRAE/IESNA 90.1 in its entirety.

> **Exception:** Buildings conforming to Section 506, provided Sections 502.4, 503.2, 504, 505.2, 505.3, 505.4, 505.6 and 505.7 are each satisfied.

SECTION 502
BUILDING ENVELOPE REQUIREMENTS

502.1 General (Prescriptive).

502.1.1 Insulation and fenestration criteria. The *building thermal envelope* shall meet the requirements of Tables 502.2(1) and 502.3 based on the climate *zone* specified in Chapter 3. Commercial buildings or portions of commercial buildings enclosing Group R occupancies shall use the *R*-values from the "Group R" column of Table 502.2(1). Commercial buildings or portions of commercial buildings enclosing occupancies other than Group R shall use the *R*-values from the "All other" column of Table 502.2(1). Buildings with a vertical fenestration area or skylight area that exceeds that allowed in Table 502.3 shall comply with the building envelope provisions of ASHRAE/IESNA 90.1.

502.1.2 *U*-factor alternative. An assembly with a *U*-factor, *C*-factor, or *F*-factor equal or less than that specified in Table 502.1.2 shall be permitted as an alternative to the *R*-value in Table 502.2(1). Commercial buildings or portions of commercial buildings enclosing Group R occupancies shall use the *U*-factor, *C*-factor, or *F*-factor from the "Group R" column of Table 502.1.2. Commercial buildings or portions of commercial buildings enclosing occupancies other than Group R shall use the *U*-factor, *C*-factor or *F*-factor from the "All other" column of Table 502.1.2.

502.2 Specific insulation requirements (Prescriptive). Opaque assemblies shall comply with Table 502.2(1).

502.2.1 Roof assembly. The minimum thermal resistance (*R*-value) of the insulating material installed either between the roof framing or continuously on the roof assembly shall be as specified in Table 502.2(1), based on construction materials used in the roof assembly.

> **Exception:** Continuously insulated roof assemblies where the thickness of insulation varies 1 inch (25 mm) or less and where the area-weighted *U*-factor is equivalent to the same assembly with the *R*-value specified in Table 502.2(1).

Insulation installed on a suspended ceiling with removable ceiling tiles shall not be considered part of the minimum thermal resistance of the roof insulation.

502.2.2 Classification of walls. Walls associated with the building envelope shall be classified in accordance with Section 502.2.2.1 or 502.2.2.2.

502.2.2.1 Above-grade walls. Above-grade walls are those walls covered by Section 502.2.3 on the exterior of the building and completely above grade or walls that are more than 15 percent above grade.

502.2.2.2 Below-grade walls. Below-grade walls covered by Section 502.2.4 are basement or first-story walls associated with the exterior of the building that are at least 85 percent below grade.

502.2.3 Above-grade walls. The minimum thermal resistance (*R*-value) of the insulating material(s) installed in the wall cavity between the framing members and continuously on the walls shall be as specified in Table 502.2(1), based on framing type and construction materials used in the wall assembly. The *R*-value of integral insulation installed in concrete masonry units (CMU) shall not be used in determining compliance with Table 502.2(1). "Mass walls" shall include walls weighing at least (1) 35 pounds per square foot (170 kg/m²) of wall surface area or (2) 25 pounds per square foot (120 kg/m²) of wall surface area if the material weight is not more than 120 pounds per cubic foot (1900 kg/m³).

502.2.4 Below-grade walls. The minimum thermal resistance (*R*-value) of the insulating material installed in, or continuously on, the below-grade walls shall be as specified in Table 502.2(1), and shall extend to a depth of 10 feet (3048 mm) below the outside finished ground level, or to the level of the floor, whichever is less.

502.2.5 Floors over outdoor air or unconditioned space. The minimum thermal resistance (*R*-value) of the insulating material installed either between the floor framing or continuously on the floor assembly shall be as specified in Table 502.2(1), based on construction materials used in the floor assembly.

"Mass floors" shall include floors weighing at least (1) 35 pounds per square foot (170 kg/m²) of floor surface area or (2) 25 pounds per square foot (120 kg/m²) of floor surface area if the material weight is not more than 120 pounds per cubic foot (1,900 kg/m³).

TABLE 502.1.2
BUILDING ENVELOPE REQUIREMENTS OPAQUE ELEMENT, MAXIMUM *U*-FACTORS

CLIMATE ZONE	1 All other	1 Group R	2 All other	2 Group R	3 All other	3 Group R	4 EXCEPT MARINE All other	4 Group R	5 AND MARINE 4 All other	5 Group R	6 All other	6 Group R	7 All other	7 Group R	8 All other	8 Group R
Roofs																
Insulation entirely above deck	U-0.063	U-0.048	U-0.048	U-0.048	U-0.048	U-0.048	U-0.048	U-0.048	U-0.048	U-0.048	U-0.048	U-0.048	U-0.039	U-0.039	U-0.039	U-0.039
Metal buildings	U-0.065	U-0.065	U-0.055	U-0.055	U-0.055	U-0.055	U-0.055	U-0.055	U-0.055	U-0.055	U-0.049	U-0.049	U-0.049	U-0.049	U-0.035	U-0.035
Attic and other	U-0.034	U-0.027	U-0.027	U-0.027	U-0.027	U-0.027	U-0.027	U-0.027	U-0.027	U-0.027	U-0.027	U-0.027	U-0.027	U-0.027	U-0.027	U-0.027
Walls, Above Grade																
Mass	U-0.58	U-0.151	U-0.151	U-0.123	U-0.123	U-0.104	U-0.104	U-0.090	U-0.090	U-0.080	U-0.080	U-0.071	U-0.071	U-0.071	U-0.071	U-0.052
Metal building	U-0.093	U-0.093	U-0.093	U-0.093	U-0.084	U-0.084	U-0.084	U-0.084	U-0.069	U-0.069	U-0.069	U-0.069	U-0.057	U-0.057	U-0.057	U-0.057
Metal framed	U-0.124	U-0.124	U-0.124	U-0.064	U-0.084	U-0.064	U-0.064	U-0.064	U-0.064	U-0.064	U-0.064	U-0.057	U-0.064	U-0.052	U-0.064	U-0.037
Wood framed and other	U-0.089	U-0.089	U-0.089	U-0.089	U-0.089	U-0.089	U-0.089	U-0.064	U-0.064	U-0.051	U-0.051	U-0.051	U-0.051	U-0.051	U-0.036	U-0.036
Walls, Below Grade																
Below-grade wall[a]	C-1.140	C-1.140	C-1.140	C-1.140	C-1.140	C-1.140	C-1.140	C-0.119	C-0.119	C-0.119	C-0.119	C-0.119	C-0.119	C-0.092	C-0.119	C-0.075
Floors																
Mass	U-0.322	U-0.322	U-0.107	U-0.087	U-0.107	U-0.087	U-0.087	U-0.074	U-0.074	U-0.074	U-0.064	U-0.057	U-0.064	U-0.051	U-0.051	U-0.051
Joist/Framing	U-0.282	U-0.282	U-0.052	U-0.052	—	U-0.033	U-0.033	U-0.033	U-0.033	U-0.033	U-0.033	U-0.033	U-0.033	U-0.033	U-0.033	U-0.033
Slab-on-Grade Floors																
Unheated slabs	F-0.730	F-0.730	F-0.730	F-0.730	F-0.730	F-0.730	F-0.730	F-0.730	F-0.730	F-0.540	F-0.540	F-0.520	F-0.520	F-0.520	F-0.520	F-0.510
Heated slabs	F-1.020	F-1.020	F-1.020	F-1.020	F-0.900	F-0.900	F-0.860	F-0.860	F-0.860	F-0.860	F-0.860	F-0.688	F-0.830	F-0.688	F-0.688	F-0.688

a. When heated slabs are placed below-grade, below grade walls must meet the *F*-factor requirements for perimeter insulation according to the heated slab-on-grade construction.

TABLE 502.2(1)
BUILDING ENVELOPE REQUIREMENTS - OPAQUE ASSEMBLIES

CLIMATE ZONE	1		2		3		4 EXCEPT MARINE		5 AND MARINE 4		6		7		8	
	All other	Group R	All other	Group R	All other	Group R	All other	Group R	All other	Group R	All other	Group R	All other	Group R	All other	Group R
Roofs																
Insulation entirely above deck	R-15ci	R-20ci	R-20ci	R-20ci	R-20ci	R-20ci	R-20ci	R-20ci	R-20ci	R-20ci	R-20ci	R-20ci	R-25ci	R-25ci	R-25ci	R-25ci
Metal buildings (with R-5 thermal blocks[a,b])	R-19	R-19	R-13 + R-13	R-13 + R-13	R-13 + R-13	R-13 + R-13	R-13 + R-13	R-13 + R-13	R-13 + R-13	R-19	R-13 + R-19	R-19	R-13 + R-19	R-19 + R-10	R-11 + R-19	R-19 + R-10
Attic and other	R-30	R-38	R-38	R-38	R-38	R-38	R-38	R-38	R-38	R-38	R-38	R-38	R-49	R-38	R-49	R-49
Walls, Above Grade																
Mass	NR	R-5.7ci[c]	R-5.7ci[c]	R-7.6ci	R-7.6ci	R-9.5ci	R-9.5ci	R-11.4ci	R-11.4ci	R-13.3ci	R-13.3ci	R-15.2ci	R-15.2ci	R-15.2ci	R-25ci	R-25ci
Metal building[b]	R-16	R-16	R-16	R-16	R-19	R-19	R-19	R-19	R-13 + R-5.6ci	R-13 + R-5.6ci	R-13 + R-5.6ci	R-13 + R-5.6ci	R-19 + R-5.6ci	R-19 + R-5.6ci	R-19 + R-5.6ci	R-19 + R-5.6ci
Metal framed	R-13	R-13	R-13	R-13+ 7.5ci	R-13 + R-3.8ci	R-13 + R-7.5ci	R-13 + R-7.5ci	R-13 + R-7.5ci	R-13 + R-7.5 ci	R-13 + R-7.5ci	R-13 + R-7.5ci	R-13 + R-7.5ci	R-13+ R-7.5ci	R-13 + R-15.6ci	R-13 + R-7.5 ci	R-13 + R-18.8ci
Wood framed and other	R-13	R-13	R-13	R-13	R-13	R-13	R-13	R-13+ R-3.8ci	R-13 - R-3.8ci	R-13 + R-3.8ci	R-13 + R-7.5ci	R-13 + R-7.5ci	R-13+ R-7.5ci	R-13 +7.5ci	R-13 + R-15.6ci	R-13 + 15.6ci
Walls, Below Grade																
Below grade wall[d]	NR	NR	NR	NR	NR	R-7.5ci	NR	R-7.5ci	R-7.5ci	R-7.5ci	R-7.5ci	R-7.5ci	R-7.5ci	R-10ci	R-7.5ci	R-12.5ci
Floors																
Mass	NR	NR	R-6.3ci	R-8.3ci	R-6.3ci	R-8.3ci	R-10ci	R-10.4ci	R-10ci	R-12.5ci	R-12.5ci	R-14.6ci	R-15ci	R-16.7ci	R-16.7ci	R-16.7ci
Joist/Framing (steel/wood)	NR	NR	R-19	R-30	R-19	R-30	R-30	R-30[e]	R-30	R-30[e]	R-30	R-30[e]	R-30	R-30[e]	R-30[e]	R-30[e]
Slab-on-Grade Floors																
Unheated slabs	NR	NR	NR	NR	NR	NR	NR	NR	NR	NR	R-10 for 24 in. below	R-15 for 24 in. below	R-15 for 24 in. below	R-15 for 24 in. below	R-15 for 24 in. below	R-20 for 24 in. below
Heated slabs	R-7.5 for 12 in. below	R-7.5 for 12 in. below	R-7.5 for 12 in. below	R-7.5 for 12 in. below	R-10 for 24 in. below	R-10 for 24 in. below	R-15 for 24 in. below	R-15 for 24 in. below	R-15 for 24 in. below	R-15 for 24 in. below	R-15 for 24 in. below	R-20 for 48 in. below	R-20 for 24 in. below	R-20 for 48 in. below	R-20 for 48 in. below	R-20 for 48 in. below
Opaque doors																
Swinging	U-0.70	U-0.70	U-0.70	U-0.70	U-0.70	U-0.70	U-0.70	U-0.70	U-0.70	U-0.70	U-0.70	U-0.70	U-0.50	U-0.50	U-0.50	U-0.50
Roll-up or sliding	U-1.45	U-1.45	U-1.45	U-1.45	U-1.45	U-1.45	U-0.50	U-0.50	U-0.50	U-0.50	U-0.50	U-0.50	U-0.50	U-0.50	U-0.50	U-0.50

For SI: 1 inch = 25.4 mm.

ci = Continuous insulation. NR = No requirement.

a. When using R-value compliance method, a thermal spacer block is required, otherwise use the U-factor compliance method. [see Tables 502.1.2 and 502.2(2)].

b. Assembly descriptions can be found in Table 502.2(2).

c. R-5.7 ci is allowed to be substituted with concrete block walls complying with ASTM C 90, ungrouted or partially grouted at 32 inches or less on center vertically and 48 inches or less on center horizontally, with ungrouted cores filled with material having a maximum thermal conductivity of 0.44 Btu-in./hr · ft² · °F.

d. When heated slabs are placed below grade, below-grade walls must meet the exterior insulation requirements for perimeter insulation according to the heated slab-on-grade construction.

e. Steel floor joist systems shall be R-38.

TABLE 502.2(2)
BUILDING ENVELOPE REQUIREMENTS–OPAQUE ASSEMBLIES

ROOFS	DESCRIPTION	REFERENCE
R-19	Standing seam roof with single fiberglass insulation layer. This construction is R-19 faced fiberglass insulation batts draped perpendicular over the purlins. A minimum R-3.5 thermal spacer block is placed above the purlin/batt, and the roof deck is secured to the purlins.	ANSI/ASHRAE/IESNA 90.1 Table A2.3 including Addendum "G"
R-13 + R-13 R-13 + R-19	Standing seam roof with two fiberglass insulation layers. The first R-value is for faced fiberglass insulation batts draped over purlins. The second R-value is for unfaced fiberglass insulation batts installed parallel to the purlins. A minimum R-3.5 thermal spacer block is placed above the purlin/batt, and the roof deck is secured to the purlins.	ANSI/ASHRAE/IESNA 90.1 Table A2.3 including Addendum "G"
R-11 + R-19 FC	Filled cavity fiberglass insulation. A continuous vapor barrier is installed below the purlins and uninterrupted by framing members. Both layers of uncompressed, unfaced fiberglass insulation rest on top of the vapor barrier and are installed parallel, between the purlins. A minimum R-3.5 thermal spacer block is placed above the purlin/batt, and the roof deck is secured to the purlins.	ANSI/ASHRAE/IESNA 90.1 Table A2.3 including Addendum "G"
WALLS		
R-16, R-19	Single fiberglass insulation layer. The construction is faced fiberglass insulation batts installed vertically and compressed between the metal wall panels and the steel framing.	ANSI/ASHRAE/IESNA 90.1 Table A3.2 including Addendum "G"
R-13 + R-5.6 ci R-19 + R-5.6 ci	The first R-value is for faced fiberglass insulation batts installed perpendicular and compressed between the metal wall panels and the steel framing. The second rated R-value is for continuous rigid insulation installed between the metal wall panel and steel framing, or on the interior of the steel framing.	ANSI/ASHRAE/IESNA 90.1 Table A3.2 including Addendum "G"

502.2.6 Slabs on grade. The minimum thermal resistance (R-value) of the insulation around the perimeter of unheated or heated slab-on-grade floors shall be as specified in Table 502.2(1). The insulation shall be placed on the outside of the foundation or on the inside of a foundation wall. The insulation shall extend downward from the top of the slab for a minimum distance as shown in the table or to the top of the footing, whichever is less, or downward to at least the bottom of the slab and then horizontally to the interior or exterior for the total distance shown in the table.

502.2.7 Opaque doors. Opaque doors (doors having less than 50 percent glass area) shall meet the applicable requirements for doors as specified in Table 502.2(1) and be considered as part of the gross area of above-grade walls that are part of the building envelope.

502.3 Fenestration (Prescriptive). Fenestration shall comply with Table 502.3.

502.3.1 Maximum area. The vertical fenestration area (not including opaque doors) shall not exceed the percentage of the gross wall area specified in Table 502.3. The skylight area shall not exceed the percentage of the gross roof area specified in Table 502.3.

502.3.2 Maximum U-factor and SHGC. For vertical fenestration, the maximum U-factor and solar heat gain coefficient (SHGC) shall be as specified in Table 502.3, based on the window projection factor. For skylights, the maximum U-factor and solar heat gain coefficient (SHGC) shall be as specified in Table 502.3.

The window projection factor shall be determined in accordance with Equation 5-1.

$$PF = A/B \qquad \textbf{(Equation 5-1)}$$

where:

PF = Projection factor (decimal).

A = Distance measured horizontally from the furthest continuous extremity of any overhang, eave, or permanently attached shading device to the vertical surface of the glazing.

B = Distance measured vertically from the bottom of the glazing to the underside of the overhang, eave, or permanently attached shading device.

Where different windows or glass doors have different PF values, they shall each be evaluated separately, or an area-weighted PF value shall be calculated and used for all windows and glass doors.

502.4 Air leakage (Mandatory).

502.4.1 Window and door assemblies. The air leakage of window and sliding or swinging door assemblies that are part of the building envelope shall be determined in accordance with AAMA/WDMA/CSA 101/I.S.2/A440, or NFRC 400 by an accredited, independent laboratory, and *labeled* and certified by the manufacturer and shall not exceed the values in Section 402.4.4.

> **Exception:** Site-constructed windows and doors that are weatherstripped or sealed in accordance with Section 502.4.3.

502.4.2 Curtain wall, storefront glazing and commercial entrance doors. Curtain wall, *storefront* glazing and commercial-glazed swinging entrance doors and revolving doors shall be tested for air leakage at 1.57 pounds per square foot (psf) (75 Pa) in accordance with ASTM E 283. For curtain walls and *storefront* glazing, the maximum air leakage rate shall be 0.3 cubic foot per minute per square foot (cfm/ft^2) (5.5 m^3/h × m^2) of fenestration area. For commercial glazed swinging entrance doors and revolving doors, the maximum air leakage rate shall be 1.00 cfm/ft^2 (18.3 m^3/h × m^2) of door area when tested in accordance with ASTM E 283.

502.4.3 Sealing of the building envelope. Openings and penetrations in the building envelope shall be sealed with caulking materials or closed with gasketing systems compatible with the construction materials and location. Joints and seams shall be sealed in the same manner or taped or covered with a moisture vapor-permeable wrapping material. Sealing materials spanning joints between construction materials shall allow for expansion and contraction of the construction materials.

502.4.5 Outdoor air intakes and exhaust openings. Stair and elevator shaft vents and other outdoor air intakes and exhaust openings integral to the building envelope shall be equipped with not less than a Class I motorized, leakage-rated damper with a maximum leakage rate of 4 cfm per square foot (6.8 L/s · C m^2) at 1.0 inch water gauge (w.g.) (1250 Pa) when tested in accordance with AMCA 500D.

> **Exception:** Gravity (nonmotorized) dampers are permitted to be used in buildings less than three stories in height above grade.

502.4.6 Loading dock weather-seals. Cargo doors and loading dock doors shall be equipped with weather-seals to restrict infiltration when vehicles are parked in the doorway.

502.4.7 Vestibules. A door that separates *conditioned space* from the exterior shall be protected with an enclosed vestibule, with all doors opening into and out of the vestibule equipped with self-closing devices. Vestibules shall be designed so that in passing through the vestibule it is not necessary for the interior and exterior doors to open at the same time.

> **Exceptions:**
> 1. Buildings in climate Zones 1 and 2 as indicated in Figure 301.1 and Table 301.1.

TABLE 502.3
BUILDING ENVELOPE REQUIREMENTS: FENESTRATION

CLIMATE ZONE	1	2	3	4 EXCEPT MARINE	5 AND MARINE 4	6	7	8
Vertical fenestration (40% maximum of above-grade wall)								
***U*-factor**								
Framing materials other than metal with or without metal reinforcement or cladding								
U-factor	1.20	0.75	0.65	0.40	0.35	0.35	0.35	0.35
Metal framing with or without thermal break								
Curtain wall/storefront *U*-factor	1.20	0.70	0.60	0.50	0.45	0.45	0.40	0.40
Entrance door *U*-factor	1.20	1.10	0.90	0.85	0.80	0.80	0.80	0.80
All other *U*-factor[a]	1.20	0.75	0.65	0.55	0.55	0.55	0.45	0.45
SHGC-all frame types								
SHGC: PF < 0.25	0.25	0.25	0.25	0.40	0.40	0.40	0.45	0.45
SHGC: 0.25 ≤ PF < 0.5	0.33	0.33	0.33	NR	NR	NR	NR	NR
SHGC: PF ≥ 0.5	0.40	0.40	0.40	NR	NR	NR	NR	NR
Skylights (3% maximum)								
U-factor	0.75	0.75	0.65	0.60	0.60	0.60	0.60	0.60
SHGC	0.35	0.35	0.35	0.40	0.40	0.40	NR	NR

NR = No requirement.

PF = Projection factor (see Section 502.3.2).

a. All others includes operable windows, fixed windows and nonentrance doors.

2. Doors not intended to be used as a building *entrance door*, such as doors to mechanical or electrical equipment rooms.

3. Doors opening directly from a *sleeping unit* or dwelling unit.

4. Doors that open directly from a space less than 3,000 square feet (298 m²) in area.

5. Revolving doors.

6. Doors used primarily to facilitate vehicular movement or material handling and adjacent personnel doors.

502.4.8 Recessed lighting. Recessed luminaires installed in the *building thermal envelope* shall be sealed to limit air leakage between conditioned and unconditioned spaces. All recessed luminaires shall be IC-rated and *labeled* as meeting ASTM E 283 when tested at 1.57 psf (75 Pa) pressure differential with no more than 2.0 cfm (0.944 L/s) of air movement from the *conditioned space* to the ceiling cavity. All recessed luminaires shall be sealed with a gasket or caulk between the housing and interior wall or ceiling covering.

SECTION 503
BUILDING MECHANICAL SYSTEMS

503.1 General. Mechanical systems and equipment serving the building heating, cooling or ventilating needs shall comply with Section 503.2 (referred to as the mandatory provisions) and either:

1. Section 503.3 (Simple systems), or

2. Section 503.4 (Complex systems).

503.2 Provisions applicable to all mechanical systems (Mandatory).

503.2.1 Calculation of heating and cooling loads. Design loads shall be determined in accordance with the procedures described in the ASHRAE/ACCA Standard 183. Heating and cooling loads shall be adjusted to account for load reductions that are achieved when energy recovery systems are utilized in the HVAC system in accordance with the ASHRAE *HVAC Systems and Equipment Handbook*. Alternatively, design loads shall be determined by an *approved* equivalent computation procedure, using the design parameters specified in Chapter 3.

503.2.2 Equipment and system sizing. Equipment and system sizing. Heating and cooling equipment and systems capacity shall not exceed the loads calculated in accordance with Section 503.2.1. A single piece of equipment providing both heating and cooling must satisfy this provision for one function with the capacity for the other function as small as possible, within available equipment options.

Exceptions:

1. Required standby equipment and systems provided with controls and devices that allow such systems or equipment to operate automatically only when the primary equipment is not operating.

2. Multiple units of the same equipment type with combined capacities exceeding the design load and provided with controls that have the capability to sequence the operation of each unit based on load.

503.2.3 HVAC equipment performance requirements. Equipment shall meet the minimum efficiency requirements of Tables 503.2.3(1), 503.2.3(2), 503.2.3(3), 503.2.3(4), 503.2.3(5), 503.2.3(6), 503.2.3(7) and 503.2.3(8) when tested and rated in accordance with the applicable test procedure. The efficiency shall be verified through certification under an *approved* certification program or, if no certification program exists, the equipment efficiency ratings shall be supported by data furnished by the manufacturer. Where multiple rating conditions or performance requirements are provided, the equipment shall satisfy all stated requirements. Where components, such as indoor or outdoor coils, from different manufacturers are used, calculations and supporting data shall be furnished by the designer that demonstrates that the combined efficiency of the specified components meets the requirements herein.

Exception: Water-cooled centrifugal water-chilling packages listed in Table 503.2.3(7) not designed for operation at ARHI Standard 550/590 test conditions of 44°F (7°C) leaving chilled water temperature and 85°F (29°C) entering condenser water temperature with 3 gpm/ton (0.054 I/s.kW) condenser water flow shall have maximum full load and NPLV ratings adjusted using the following equations:

Adjusted maximum full load kW/ton rating = [full load kW/ton from Table 503.2.3(7)]/K_{adj}

Adjusted maximum NPLV rating = [IPLV from Table 503.2.3(7)]/K_{adj}

where:

K_{adj} = $6.174722 - 0.303668(X) + 0.00629466(X)^2 - 0.000045780(X)^3$

X = DT_{std} + LIFT

DT_{std} = {24+[full load kW/ton from Table 503.2.3(7)] × 6.83}/Flow

Flow = Condenser water flow (GPM)/Cooling Full Load Capacity (tons)

LIFT = CEWT – CLWT (°F)

CEWT = Full Load Condenser Entering Water Temperature (°F)

CLWT = Full Load Leaving Chilled Water Temperature (°F)

The adjusted full load and NPLV values are only applicable over the following full-load design ranges:

Minimum Leaving Chilled
Water Temperature: 38°F (3.3°C)

Maximum Condenser Entering
Water Temperature:　　　102°F (38.9°C)

Condensing Water Flow: 1 to 6 gpm/ton 0.018 to
0.1076 1/s · kW) and X ≥ 39 and ≤ 60

Chillers designed to operate outside of these ranges or applications utilizing fluids or solutions with secondary coolants (e.g., glycol solutions or brines) with a freeze point of 27°F (-2.8°C) or lower for freeze protection are not covered by this code.

TABLE 503.2.3(1)
UNITARY AIR CONDITIONERS AND CONDENSING UNITS, ELECTRICALLY OPERATED, MINIMUM EFFICIENCY REQUIREMENTS

EQUIPMENT TYPE	SIZE CATEGORY	SUBCATEGORY OR RATING CONDITION	MINIMUM EFFICIENCY[b]	TEST PROCEDURE[a]
Air conditioners, Air cooled	< 65,000 Btu/h[d]	Split system	13.0 SEER	AHRI 210/240
		Single package	13.0 SEER	
	≥ 65,000 Btu/h and < 135,000 Btu/h	Split system and single package	10.3 EER[c] (before Jan 1, 2010) 11.2 EER[c] (as of Jan 1, 2010)	
	≥ 135,000 Btu/h and < 240,000 Btu/h	Split system and single package	9.7 EER[c] (before Jan 1, 2010) 11.0 EER[c] (as of Jan 1, 2010)	AHRI 340/360
	≥ 240,000 Btu/h and < 760,000 Btu/h	Split system and single package	9.5 EER[c] 9.7 IPLV[c] (before Jan 1, 2010) 10.0 EER[c] 9.7 IPLV[g] (as of Jan 1, 2010)	
	≥ 760,000 Btu/h	Split system and single package	9.2 EER[c] 9.4 IPLV[c] (before Jan 1, 2010) 9.7 EER[c] 9.4 IPLV[c] (as of Jan 1, 2010)	
Through-the-wall, Air cooled	< 30,000 Btu/h[d]	Split system	10.9 SEER (before Jan 23, 2010) 12.0 SEER (as of Jan 23, 2010)	AHRI 210/240
		Single package	10.6 SEER (before Jan 23, 2010) 12.0 SEER (as of Jan 23, 2010)	
Air conditioners, Water and evaporatively cooled	< 65,000 Btu/h	Split system and single package	12.1 EER	AHRI 210/240
	≥ 65,000 Btu/h and < 135,000 Btu/h	Split system and single package	11.5 EER[c]	
	≥ 135,000 Btu/h and < 240,000 Btu/h	Split system and single package	11.0 EER[c]	AHRI 340/360
	≥ 240,000 Btu/h	Split system and single package	11.5 EER[c]	

For SI:　1 British thermal unit per hour = 0.2931 W.

a. Chapter 6 contains a complete specification of the referenced test procedure, including the referenced year version of the test procedure.

b. IPLVs are only applicable to equipment with capacity modulation.

c. Deduct 0.2 from the required EERs and IPLVs for units with a heating section other than electric resistance heat.

d. Single-phase air-cooled air conditioners < 65,000 Btu/h are regulated by the National Appliance Energy Conservation Act of 1987 (NAECA); SEER values are those set by NAECA.

TABLE 503.2.3(2)
UNITARY AIR CONDITIONERS AND CONDENSING UNITS, ELECTRICALLY OPERATED, MINIMUM EFFICIENCY REQUIREMENTS

EQUIPMENT TYPE	SIZE CATEGORY	SUBCATEGORY OR RATING CONDITION	MINIMUM EFFICIENCY[b]	TEST PROCEDURE[a]
Air cooled, (Cooling mode)	< 65,000 Btu/h[d]	Split system	13.0 SEER	AHRI 210/240
		Single package	13.0 SEER	
	≥ 65,000 Btu/h and < 135,000 Btu/h	Split system and single package	10.1 EER[c] (before Jan 1, 2010) 11.0 EER[c] (as of Jan 1, 2010)	
	≥ 135,000 Btu/h and < 240,000 Btu/h	Split system and single package	9.3 EER[c] (before Jan 1, 2010) 10.6 EER[c] (as of Jan 1, 2010)	AHRI 340/360
	≥ 240,000 Btu/h	Split system and single package	9.0 EER[c] 9.2 IPLV[c] (before Jan 1, 2010) 9.5 EER[c] 9.2 IPLV[c] (as of Jan 1, 2010)	
Through-the-Wall (Air cooled, cooling mode)	< 30,000 Btu/h[d]	Split system	10.9 SEER (before Jan 23, 2010) 12.0 SEER (as of Jan 23, 2010)	AHRI 210/240
		Single package	10.6 SEER (before Jan 23, 2010) 12.0 SEER (as of Jan 23, 2010)	
Water Source (Cooling mode)	< 17,000 Btu/h	86°F entering water	11.2 EER	AHRI/ASHRAE 13256-1
	≥ 17,000 Btu/h and < 135,000 Btu/h	86°F entering water	12.0 EER	AHRIASHRAE 13256-1
Groundwater Source (Cooling mode)	< 135,000 Btu/h	59°F entering water	16.2 EER	AHRI/ASHRAE 13256-1
Ground source (Cooling mode)	< 135,000 Btu/h	77°F entering water	13.4 EER	AHRI/ASHRAE 13256-1
Air cooled (Heating mode)	< 65,000 Btu/h[d] (Cooling capacity)	Split system	7.7 HSPF	AHRI 210/240
		Single package	7.7 HSPF	
	≥ 65,000 Btu/h and < 135,000 Btu/h (Cooling capacity)	47°F db/43°F wb Outdoor air	3.2 COP (before Jan 1, 2010) 3.3 COP (as of Jan 1, 2010)	
	≥ 135,000 Btu/h (Cooling capacity)	47°F db/43°F wb Outdoor air	3.1 COP (before Jan 1, 2010) 3.2 COP (as of Jan 1, 2010)	AHRI 340/360

(continued)

TABLE 503.2.3(2)—continued
UNITARY AIR CONDITIONERS AND CONDENSING UNITS, ELECTRICALLY OPERATED, MINIMUM EFFICIENCY REQUIREMENTS

EQUIPMENT TYPE	SIZE CATEGORY	SUBCATEGORY OR RATING CONDITION	MINIMUM EFFICIENCY[b]	TEST PROCEDURE[a]
Through-the-wall (Air cooled, heating mode)	< 30,000 Btu/h	Split System	7.1 HSPE (before Jan 23, 2010) 7.4 HSPF (as of Jan 23, 2010)	AHRI 210/240
		Single package	7.0 HSPF (before Jan 23, 2010) 7.4 HSPF (as of Jan 23, 2010)	
Water source (Heating mode)	< 135,000 Btu/h (Cooling capacity)	68°F entering water	4.2 COP	AHRI/ASHRAE 13256-1
Groundwater source (Heating mode)	< 135,000 Btu/h (Cooling capacity)	50°F entering water	3.6 COP	AHRI/ASHRAE 13256-1
Ground source (Heating mode)	< 135,000 Btu/h (Cooling capacity)	32°F entering water	3.1 COP	AHRI/ASHRAE 13256-1

For SI: °C = [(°F) - 32]/1.8, 1 British thermal unit per hour = 0.2931 W.

db = dry-bulb temperature, °F; wb = wet-bulb temperature, °F.

a. Chapter 6 contains a complete specification of the referenced test procedure, including the referenced year version of the test procedure.

b. IPLVs and Part load rating conditions are only applicable to equipment with capacity modulation.

c. Deduct 0.2 from the required EERs and IPLVs for units with a heating section other than electric resistance heat.

d. Single-phase air-cooled heat pumps < 65,000 Btu/h are regulated by the National Appliance Energy Conservation Act of 1987 (NAECA), SEER and HSPF values are those set by NAECA.

TABLE 503.2.3(3)
PACKAGED TERMINAL AIR CONDITIONERS AND PACKAGED TERMINAL HEAT PUMPS

EQUIPMENT TYPE	SIZE CATEGORY (INPUT)	SUBCATEGORY OR RATING CONDITION	MINIMUM EFFICIENCY[b]	TEST PROCEDURE[a]
PTAC (Cooling mode) New construction	All capacities	95°F db outdoor air	12.5 - (0.213 · Cap/1000) EER	AHRI 310/380
PTAC (Cooling mode) Replacements[c]	All capacities	95°F db outdoor air	10.9 - (0.213 · Cap/1000) EER	
PTHP (Cooling mode) New construction	All capacities	95°F db outdoor air	12.3 - (0.213 · Cap/1000) EER	
PTHP (Cooling mode) Replacements[c]	All capacities	95°F db outdoor air	10.8 - (0.213 · Cap/1000) EER	
PTHP (Heating mode) New construction	All capacities	—	3.2 - (0.026 · Cap/1000) COP	
PTHP (Heating mode) Replacements[c]	All capacities	—	2.9 - (0.026 · Cap/1000) COP	

For SI: °C - [(°F) - 32]/1.8, 1 British thermal unit per hour - 0.2931 W.

db = dry-bulb temperature, °F.

wb = wet-bulb temperature, °F.

a. Chapter 6 contains a complete specification of the referenced test procedure, including the referenced year version of the test procedure.

b. Cap means the rated cooling capacity of the product in Btu/h. If the unit's capacity is less than 7,000 Btu/h, use 7,000 Btu/h in the calculation. If the unit's capacity is greater than 15,000 Btu/h, use 15,000 Btu/h in the calculation.

c. Replacement units must be factory labeled as follows: "MANUFACTURED FOR REPLACEMENT APPLICATIONS ONLY: NOT TO BE INSTALLED IN NEW CONSTRUCTION PROJECTS." Replacement efficiencies apply only to units with existing sleeves less than 16 inches (406 mm) high and less than 42 inches (1067 mm) wide.

TABLE 503.2.3(4)
WARM AIR FURNACES AND COMBINATION WARM AIR FURNACES/AIR-CONDITIONING UNITS,
WARM AIR DUCT FURNACES AND UNIT HEATERS, MINIMUM EFFICIENCY REQUIREMENTS

EQUIPMENT TYPE	SIZE CATEGORY (INPUT)	SUBCATEGORY OR RATING CONDITION	MINIMUM EFFICIENCY [d, e]	TEST PROCEDURE[a]
Warm air furnaces, gas fired	< 225,000 Btu/h	—	78% AFUE or 80% E_t[c]	DOE 10 CFR Part 430 or ANSI Z21.47
	≥ 225,000 Btu/h	Maximum capacity[c]	80% E_t[f]	ANSI Z21.47
Warm air furnaces, oil fired	< 225,000 Btu/h	—	78% AFUE or 80% E_t[c]	DOE 10 CFR Part 430 or UL 727
	≥ 225,000 Btu/h	Maximum capacity[b]	81% E_t[g]	UL 727
Warm air duct furnaces, gas fired	All capacities	Maximum capacity[b]	80% E_c	ANSI Z83.8
Warm air unit heaters, gas fired	All capacities	Maximum capacity[b]	80% E_c	ANSI Z83.8
Warm air unit heaters, oil fired	All capacities	Maximum capacity[b]	80% E_c	UL 731

For SI: 1 British thermal unit per hour = 0.2931 W.

a. Chapter 6 contains a complete specification of the referenced test procedure, including the referenced year version of the test procedure.

b. Minimum and maximum ratings as provided for and allowed by the unit's controls.

c. Combination units not covered by the National Appliance Energy Conservation Act of 1987 (NAECA) (3-phase power or cooling capacity greater than or equal to 65,000 Btu/h [19 kW]) shall comply with either rating.

d. E_t = Thermal efficiency. See test procedure for detailed discussion.

e. E_c = Combustion efficiency (100% less flue losses). See test procedure for detailed discussion.

f. E_c = Combustion efficiency. Units must also include an IID, have jackets not exceeding 0.75 percent of the input rating, and have either power venting or a flue damper. A vent damper is an acceptable alternative to a flue damper for those furnaces where combustion air is drawn from the conditioned space.

g. E_t = Thermal efficiency. Units must also include an IID, have jacket losses not exceeding 0.75 percent of the input rating, and have either power venting or a flue damper. A vent damper is an acceptable alternative to a flue damper for those furnaces where combustion air is drawn from the conditioned space.

TABLE 503.2.3(5)
BOILERS, GAS- AND OIL-FIRED, MINIMUM EFFICIENCY REQUIREMENTS

EQUIPMENT TYPE[f]	SIZE CATEGORY	SUBCATEGORY OR RATING CONDITION	MINIMUM EFFICIENCY[c, d, e]	TEST PROCEDURE
Boilers, Gas-fired	< 300,000 Btu/h	Hot water	80% AFUE	DOE 10 CFR Part 430
		Steam	75% AFUE	
	≥ 300,000 Btu/h and ≤ 2,500,000 Btu/h	Minimum capacity[b]	75% E_t and 80% E_c (See Note c, d)	DOE 10 CFR Part 431
	> 2,500,000 Btu/h[f]	Hot water	80% E_c (See Note c, d)	
		Steam	80% E_c (See Note c, d)	
Boilers, Oil-fired	< 300,000 Btu/h	—	80% AFUE	DOE 10 CFR Part 430
	≥ 300,000 Btu/h and ≤ 2,500,000 Btu/h	Minimum capacity[b]	78% E_t and 83% E_c (See Note c, d)	DOE 10 CFR Part 431
	> 2,500,000 Btu/h[a]	Hot water	83% E_c (See Note c, d)	
		Steam	83% E_c (See Note c, d)	
Boilers, Oil-fired (Residual)	≥ 300,000 Btu/h and ≤ 2,500,000 Btu/h	Minimum capacity[b]	78% E_t and 83% E_c (See Note c, d)	DOE 10 CFR Part 431
	> 2,500,000 Btu/h[a]	Hot water	83% E_c (See Note c, d)	
		Steam	83% E_c (See Note c, d)	

For SI: 1 British thermal unit per hour = 0.2931 W.

a. Chapter 6 contains a complete specification of the referenced test procedure, including the referenced year version of the test procedure.

b. Minimum ratings as provided for and allowed by the unit's controls.

c. E_c = Combustion efficiency (100 percent less flue losses). See reference document for detailed information.

d. E_t = Thermal efficiency. See reference document for detailed information.

e. Alternative test procedures used at the manufacturer's option are ASME PTC-4.1 for units greater than 5,000,000 Btu/h input, or ANSI Z21.13 for units greater than or equal to 300,000 Btu/h and less than or equal to 2,500,000 Btu/h input.

f. These requirements apply to boilers with rated input of 8,000,000 Btu/h or less that are not packaged boilers, and to all packaged boilers. Minimum efficiency requirements for boilers cover all capacities of packaged boilers.

TABLE 503.2.3(6)
CONDENSING UNITS, ELECTRICALLY OPERATED, MINIMUM EFFICIENCY REQUIREMENTS

EQUIPMENT TYPE	SIZE CATEGORY	MINIMUM EFFICIENCY[b]	TEST PROCEDURE[a]
Condensing units, air cooled	≥ 135,000 Btu/h	10.1 EER 11.2 IPLV	AHRI 365
Condensing units, water or evaporatively cooled	≥ 135,000 Btu/h	13.1 EER 13.1 IPLV	

For SI: 1 British thermal unit per hour = 0.2931 W.

a. Chapter 6 contains a complete specification of the referenced test procedure, including the referenced year version of the test procedure.

b. IPLVs are only applicable to equipment with capacity modulation.

TABLE 503.2.3(7)
WATER CHILLING PACKAGES, EFFICIENCY REQUIREMENTS[a]

| EQUIPMENT TYPE | SIZE CATEGORY | UNITS | BEFORE 1/1/2010 | | AS OF 1/1/2010[c] | | | | TEST PROCEDURE[b] |
| | | | | | PATH A | | PATH B | | |
			FULL LOAD	IPLV	FULL LOAD	IPLV	FULL LOAD	IPLV	
Air-cooled chillers	< 150 tons	EER	≥ 9.562	≥ 10.416	≥ 9.562	≥ 12.500	NA[d]	NA[d]	AHRI 550/590
	≥ 150 tons	EER			≥ 9.562	≥ 12.750	NA[d]	NA[d]	
Air cooled without condenser, electrical operated	All capacities	EER	≥ 10.586	≥ 11.782	Air-cooled chillers without condensers must be rated with matching condensers and comply with the air-cooled chiller efficiency requirements				
Water cooled, electrically operated, reciprocating	All capacities	kW/ton	≤ 0.837	≤ 0.696	Reciprocating units must comply with water cooled positive displacement efficiency requirements				
Water cooled, electrically operated, positive displacement	< 75 tons	kW/ton	≤ 0.790	≤ 0.676	≤ 0.780	≤ 0.630	≤ 0.800	≤ 0.600	
	≥ 75 tons and < 150 tons	kW/ton			≤ 0.775	≤ 0.615	≤ 0.790	≤ 0.586	
	≥ 150 tons and < 300 tons	kW/ton	≤ 0.717	≤ 0.627	≤ 0.680	≤ 0.580	≤ 0.718	≤ 0.540	
	≥ 300 tons	kW/ton	≤ 0.639	≤ 0.571	≤ 0.620	≤ 0.540	≤ 0.639	≤ 0.490	
Water cooled, electrically operated, centrifugal	< 150 tons	kW/ton	≤ 0.703	≤ 0.669	≤ 0.634	≤ 0.596	≤ 0.639	≤ 0.450	
	≥ 150 tons and < 300 tons	kW/ton	≤ 0.634	≤ 0.596					
	≥ 300 tons and < 600 tons	kW/ton	≤ 0.576	≤ 0.549	≤ 0.576	≤ 0.549	≤ 0.600	≤ 0.400	
	≥ 600 tons	kW/ton	≤ 0.576	≤ 0.549	≤ 0.570	≤ 0.539	≤ 0.590	≤ 0.400	
Air cooled, absorption single effect	All capacities	COP	≥ 0.600	NR[e]	≥ 0.600	NR[e]	NA[d]	NA[d]	AHRI 560
Water-cooled, absorption single effect	All capacities	COP	≥ 0.700	NR[e]	≥ 0.700	NR[e]	NA[d]	NA[d]	
Absorption double effect, indirect-fired	All capacities	COP	≥ 1.000	≥ 1.050	≥ 1.000	≥ 1.050	NA[d]	NA[d]	
Absorption double effect, direct fired	All capacities	COP	≥ 1.000	≥ 1.000	≥ 1.000	≥ 1.000	NA[d]	NA[d]	

For SI: 1 ton = 3517 W, 1 British thermal unit per hour = 0.2931 W.

a. The chiller equipment requirements do not apply for chillers used in low-temperature applications where the design leaving fluid temperature is < 40°F.

b. Section 12 contains a complete specification of the referenced test procedure, including the referenced year version of the test procedure.

c. Compliance with this standard can be obtained by meeting the minimum requirements of Path A or B. However, both the full load and IPLV must be met to fulfill the requirements of Path A or B.

d. NA means that this requirement is not applicable and cannot be used for compliance.

e. NR means that there are no minimum requirements for this category.

503.2.4 HVAC system controls. Each heating and cooling system shall be provided with thermostatic controls as required in Section 503.2.4.1, 503.2.4.2, 503.2.4.3, 503.2.4.4, 503.4.1, 503.4.2, 503.4.3 or 503.4.4.

503.2.4.1 Thermostatic controls. The supply of heating and cooling energy to each zone shall be controlled by individual thermostatic controls capable of responding to temperature within the zone. Where humidification or dehumidification or both is provided, at least one humidity control device shall be provided for each humidity control system.

Exception: Independent perimeter systems that are designed to offset only building envelope heat losses or gains or both serving one or more perimeter zones also served by an interior system provided:

1. The perimeter system includes at least one thermostatic control zone for each building exposure having exterior walls facing only one orientation (within +/- 45 degrees) (0.8 rad) for more than 50 contiguous feet (15.2 m); and

2. The perimeter system heating and cooling supply is controlled by a thermostat(s) located within the zone(s) served by the system.

503.2.4.1.1 Heat pump supplementary heat. Heat pumps having supplementary electric resistance heat shall have controls that, except during defrost, prevent supplementary heat operation when the heat pump can meet the heating load.

503.2.4.2 Setpoint overlap restriction. Where used to control both heating and cooling, *zone* thermostatic controls shall provide a temperature range or dead band of at least 5°F (2.8°C) within which the supply of heating and

cooling energy to the zone is capable of being shut off or reduced to a minimum.

Exception: Thermostats requiring manual change-over between heating and cooling modes.

503.2.4.3 Off-hour controls. Each zone shall be provided with thermostatic setback controls that are controlled by either an automatic time clock or programmable control system.

Exceptions:

1. Zones that will be operated continuously.

2. Zones with a full HVAC load demand not exceeding 6,800 Btu/h (2 kW) and having a readily accessible manual shutoff switch.

503.2.4.3.1 Thermostatic setback capabilities. Thermostatic setback controls shall have the capability to set back or temporarily operate the system to maintain zone temperatures down to 55°F (13°C) or up to 85°F (29°C).

503.2.4.3.2 Automatic setback and shutdown capabilities. Automatic time clock or programmable controls shall be capable of starting and stopping the system for seven different daily schedules per week and retaining their programming and time setting during a loss of power for at least 10 hours. Additionally, the controls shall have a manual override that allows temporary operation of the system for up to 2 hours; a manually operated timer capable of being adjusted to operate the system for up to 2 hours; or an occupancy sensor.

503.2.4.4 Shutoff damper controls. Both outdoor air supply and exhaust ducts shall be equipped with motor-

TABLE 503.2.3(8)
PERFORMANCE REQUIREMENTS FOR HEAT REJECTION EQUIPMENT

EQUIPMENT TYPE	TOTAL SYSTEM HEAT REJECTION CAPACITY AT RATED CONDITIONS	SUBCATEGORY OR RATING CONDITION	PERFORMANCE REQUIRED[a,b,]	TEST PROCEDURE[c]
Propeller or axial fan cooling towers	All	95°F entering water 85°F leaving water 75°F wb outdoor air	≥ 38.2 gpm/hp	CTI ATC-105 and CTI STD-201
Centrifugal fan cooling towers	All	95°F entering water 85°F leaving water 75°F wb outdoor air	≥ 20.0 gpm/hp	CTI ATC-105 and CTI STD-201
Air-cooled condensers	All	125°F condensing temperature R-22 test fluid 190°F entering gas temperature 15°F subcooling 95°F entering db	≥ 176,000 Btu/h · hp (69 COP)	ARI 460

For SI: °C = [(°F) - 32] / 1.8, 1 British thermal unit per hour = 0.2931 W, 1 gallon per minute per horsepower = 0.846 L/s kW.

wb = wet-bulb temperature, °F

a. For purposes of this table, cooling tower performance is defined as the maximum flow rating of the tower units (gpm) divided by the fan nameplate rated motor power units (hp).

b. For purposes of this table, air-cooled condenser performance is defined as the heat rejected from the refrigerant units (Btu/h) divided by the fan nameplate rated motor power units (hp).

c. Chapter 6 contains a complete specification of the referenced test procedure, including the referenced year version of the test procedure.

ized dampers that will automatically shut when the systems or spaces served are not in use.

Exceptions:

1. Gravity dampers shall be permitted in buildings less than three stories in height.

2. Gravity dampers shall be permitted for buildings of any height located in Climate Zones 1, 2 and 3.

3. Gravity dampers shall be permitted for outside air intake or exhaust airflows of 300 cfm (0.14 m³/s) or less.

503.2.4.5 Snow melt system controls. Snow- and ice-melting systems, supplied through energy service to the building, shall include automatic controls capable of shutting off the system when the pavement temperature is above 50°F (10°C) and no precipitation is falling and an automatic or manual control that will allow shutoff when the outdoor temperature is above 40°F (4°C) so that the potential for snow or ice accumulation is negligible.

503.2.5 Ventilation. Ventilation, either natural or mechanical, shall be provided in accordance with Chapter 4 of the *International Mechanical Code*. Where mechanical ventilation is provided, the system shall provide the capability to reduce the outdoor air supply to the minimum required by Chapter 4 of the *International Mechanical Code*.

503.2.5.1 Demand controlled ventilation. Demand control ventilation (DCV) is required for spaces larger than 500 ft² (50 m²) and with an average occupant load of 40 people per 1000 ft² (93 m²) of floor area (as established in Table 403.3 of the *International Mechanical Code*) and served by systems with one or more of the following:

1. An air-side economizer;

2. Automatic modulating control of the outdoor air damper; or

3. A design outdoor airflow greater than 3,000 cfm (1400 L/s).

Exceptions:

1. Systems with energy recovery complying with Section 503.2.6.

2. Multiple-zone systems without direct digital control of individual zones communicating with a central control panel.

3. System with a design outdoor airflow less than 1,200 cfm (600 L/s).

4. Spaces where the supply airflow rate minus any makeup or outgoing transfer air requirement is less than 1,200 cfm (600 L/s).

503.2.6 Energy recovery ventilation systems. Individual fan systems that have both a design supply air capacity of 5,000 cfm (2.36 m³/s) or greater and a minimum outside air supply of 70 percent or greater of the design supply air quantity shall have an energy recovery system that provides a change in the enthalpy of the outdoor air supply of 50 percent or more of the difference between the outdoor air and return air at design conditions. Provision shall be made to bypass or control the energy recovery system to permit cooling with outdoor air where cooling with outdoor air is required.

Exception: An energy recovery ventilation system shall not be required in any of the following conditions:

1. Where energy recovery systems are prohibited by the *International Mechanical Code*.

2. Laboratory fume hood systems that include at least one of the following features:

    2.1. Variable-air-volume hood exhaust and room supply systems capable of reducing exhaust and makeup air volume to 50 percent or less of design values.

    2.2. Direct makeup (auxiliary) air supply equal to at least 75 percent of the exhaust rate, heated no warmer than 2°F (1.1°C) below room setpoint, cooled to no cooler than 3°F (1.7°C) above room setpoint, no humidification added, and no simultaneous heating and cooling used for dehumidification control.

3. Systems serving spaces that are not cooled and are heated to less than 60°F (15.5°C).

4. Where more than 60 percent of the outdoor heating energy is provided from site-recovered or site solar energy.

5. Heating systems in climates with less than 3,600 HDD.

6. Cooling systems in climates with a 1-percent cooling design wet-bulb temperature less than 64°F (18°C).

7. Systems requiring dehumidification that employ series-style energy recovery coils wrapped around the cooling coil.

503.2.7 Duct and plenum insulation and sealing. All supply and return air ducts and plenums shall be insulated with a minimum of R-5 insulation when located in unconditioned spaces and a minimum of R-8 insulation when located outside the building. When located within a building envelope assembly, the duct or plenum shall be separated from the building exterior or unconditioned or exempt spaces by a minimum of R-8 insulation.

Exceptions:

1. When located within equipment.

2. When the design temperature difference between the interior and exterior of the duct or plenum does not exceed 15°F (8°C).

All ducts, air handlers and filter boxes shall be sealed. Joints and seams shall comply with Section 603.9 of the *International Mechanical Code*.

503.2.7.1 Duct construction. Ductwork shall be constructed and erected in accordance with the *International Mechanical Code*.

503.2.7.1.1 Low-pressure duct systems. All longitudinal and transverse joints, seams and connections of supply and return ducts operating at a static pressure less than or equal to 2 inches w.g. (500 Pa) shall be securely fastened and sealed with welds, gaskets, mastics (adhesives), mastic-plus-embedded-fabric systems or tapes installed in accordance with the manufacturer's installation instructions. Pressure classifications specific to the duct system shall be clearly indicated on the construction documents in accordance with the *International Mechanical Code*.

Exception: Continuously welded and locking-type longitudinal joints and seams on ducts operating at static pressures less than 2 inches w.g. (500 Pa) pressure classification.

503.2.7.1.2 Medium-pressure duct systems. All ducts and plenums designed to operate at a static pressure greater than 2 inches w.g. (500 Pa) but less than 3 inches w.g. (750 Pa) shall be insulated and sealed in accordance with Section 503.2.7. Pressure classifications specific to the duct system shall be clearly indicated on the construction documents in accordance with the *International Mechanical Code*.

503.2.7.1.3 High-pressure duct systems. Ducts designed to operate at static pressures in excess of 3 inches w.g. (746 Pa) shall be insulated and sealed in accordance with Section 503.2.7. In addition, ducts and plenums shall be leak-tested in accordance with the SMACNA *HVAC Air Duct Leakage Test Manual* with the rate of air leakage (*CL*) less than or equal to 6.0 as determined in accordance with Equation 5-2.

$$CL = F / P^{0.65} \qquad \textbf{(Equation 5-2)}$$

where:

F = The measured leakage rate in cfm per 100 square feet of duct surface.

P = The static pressure of the test.

Documentation shall be furnished by the designer demonstrating that representative sections totaling at least 25 percent of the duct area have been tested and that all tested sections meet the requirements of this section.

503.2.8 Piping insulation. All piping serving as part of a heating or cooling system shall be thermally insulated in accordance with Table 503.2.8.

Exceptions:

1. Factory-installed piping within HVAC equipment tested and rated in accordance with a test procedure referenced by this code.

2. Factory-installed piping within room fan-coils and unit ventilators tested and rated according to AHRI 440 (except that the sampling and variation provisions of Section 6.5 shall not apply) and 840, respectively.

3. Piping that conveys fluids that have a design operating temperature range between 55°F (13°C) and 105°F (41°C).

4. Piping that conveys fluids that have not been heated or cooled through the use of fossil fuels or electric power.

5. Runout piping not exceeding 4 feet (1219 mm) in length and 1 inch (25 mm) in diameter between the control valve and HVAC coil.

TABLE 503.2.8
MINIMUM PIPE INSULATION
(thickness in inches)

FLUID	NOMINAL PIPE DIAMETER	
	≤ 1.5″	> 1.5″
Steam	$1^1/_2$	3
Hot water	$1^1/_2$	2
Chilled water, brine or refrigerant	$1^1/_2$	$1^1/_2$

For SI: 1 inch = 25.4 mm.

a. Based on insulation having a conductivity (*k*) not exceeding 0.27 Btu per inch/h · ft² · °F.

b. For insulation with a thermal conductivity not equal to 0.27 Btu · inch/h · ft² · °F at a mean temperature of 75°F, the minimum required pipe thickness is adjusted using the following equation;

$$T = r[(1+t/r)^{K/k} - 1]$$

where:

T = Adjusted insulation thickness (in).

r = Actual pipe radius (in).

t = Insulation thickness from applicable cell in table (in).

K = New thermal conductivity at 75°F (Btu · in/hr · ft² · °F).

k = 0.27 Btu · in/hr · ft² · °F.

503.2.9 HVAC system completion. Prior to the issuance of a certificate of occupancy, the design professional shall provide evidence of system completion in accordance with Sections 503.2.9.1 through 503.2.9.3.

503.2.9.1 Air system balancing. Each supply air outlet and *zone* terminal device shall be equipped with means for air balancing in accordance with the requirements of Chapter 6 of the *International Mechanical Code*. Discharge dampers are prohibited on constant volume fans and variable volume fans with motors 10 horsepower (hp) (7.5 kW) and larger.

503.2.9.2 Hydronic system balancing. Individual hydronic heating and cooling coils shall be equipped with means for balancing and pressure test connections.

503.2.9.3 Manuals. The construction documents shall require that an operating and maintenance manual be provided to the building owner by the mechanical contractor. The manual shall include, at least, the following:

1. Equipment capacity (input and output) and required maintenance actions.

2. Equipment operation and maintenance manuals.

3. HVAC system control maintenance and calibration information, including wiring diagrams, schematics, and control sequence descriptions. Desired or field-determined setpoints shall be permanently recorded on control drawings, at control devices or, for digital control systems, in programming comments.

4. A complete written narrative of how each system is intended to operate.

503.2.10 Air system design and control. Each HVAC system having a total fan system motor nameplate horsepower (hp) exceeding 5 horsepower (hp) (3.7 kW) shall meet the provisions of Sections 503.2.10.1 through 503.2.10.2.

503.2.10.1 Allowable fan floor horsepower. Each HVAC system at fan system design conditions shall not exceed the allowable fan system motor nameplate hp (Option 1) or fan system bhp (Option 2) as shown in Table 503.2.10.1(1). This includes supply fans, return/relief fans, and fan-powered terminal units associated with systems providing heating or cooling capability.

Exceptions:

1. Hospital and laboratory systems that utilize flow control devices on exhaust and/or return to maintain space pressure relationships necessary for occupant health and safety or environmental control shall be permitted to use variable volume fan power limitation.

2. Individual exhaust fans with motor nameplate horsepower of 1 hp (0.7 kW) or less.

3. Fans exhausting air from fume hoods. (Note: If this exception is taken, no related exhaust side credits shall be taken from Table 503.2.10.1(2) and the Fume Exhaust Exception Deduction must be taken from Table 503.2.10.1(2).

TABLE 503.2.10.1(1)
FAN POWER LIMITATION

	LIMIT	CONSTANT VOLUME	VARIABLE VOLUME
Option 1: Fan system motor nameplate hp	Allowable nameplate motor hp	$hp \leq CFM_S *0.0011$	$hp \leq CFM_S *0.0015$
Option 2: Fan system bhp	Allowable fan system bhp	$bhp \leq CFM_S *0.00094 + A$	$bhp \leq CFM_S *0.0013 + A$

where:

CFM_S = The maximum design supply airflow rate to conditioned spaces served by the system in cubic feet per minute.

hp = The maximum combined motor nameplate horsepower.

Bhp = The maximum combined fan brake horsepower.

A = Sum of $[PD \times CFM_D / 4131]$.

where:

PD = Each applicable pressure drop adjustment from Table 503.2.10.1(2) in. w.c.

CFM_D = The design airflow through each applicable device from Table 503.2.10.1(2) in cubic feet per minute.

TABLE 503.2.10.1(2)
FAN POWER LIMITATION PRESSURE DROP ADJUSTMENT

DEVICE	ADJUSTMENT
Credits	
Fully ducted return and/or exhaust air systems	0.5 in w.c.
Return and/or exhaust airflow control devices	0.5 in w.c
Exhaust filters, scrubbers or other exhaust treatment.	The pressure drop of device calculated at fan system design condition.
Particulate filtration credit: MERV 9 thru 12	0.5 in w.c.
Particulate filtration credit: MERV 13 thru 15	0.9 in w.c.
Particulate filtration credit: MERV 16 and greater and electronically enhanced filters	Pressure drop calculated at 2x clean filter pressure drop at fan system design condition.
Carbon and other gas-phase air cleaners	Clean filter pressure drop at fan system design condition.
Heat recovery device	Pressure drop of device at fan system design condition.
Evaporative humidifier/cooler in series with another cooling coil	Pressure drop of device at fan system design conditions
Sound attenuation section	0.15 in w.c.
Deductions	
Fume hood exhaust exception (required if Section 503.2.10.1, Exception 3, is taken)	-1.0 in w.c.

503.2.10.2 Motor nameplate horsepower. For each fan, the selected fan motor shall be no larger than the first available motor size greater than the brake horsepower (bhp). The fan brake horsepower (bhp) shall be indicated on the design documents to allow for compliance verification by the *code official.*

Exceptions:

1. For fans less than 6 bhp, where the first available motor larger than the brake horsepower has a nameplate rating within 50 percent of the bhp, selection of the next larger nameplate motor size is allowed.

2. For fans 6 bhp and larger, where the first available motor larger than the bhp has a nameplate rating within 30 percent of the bhp, selection of the next larger nameplate motor size is allowed.

503.2.11 Heating outside a building. Systems installed to provide heat outside a building shall be radiant systems. Such heating systems shall be controlled by an occupancy sensing device or a timer switch, so that the system is automatically deenergized when no occupants are present.

503.3 Simple HVAC systems and equipment (Prescriptive). This section applies to buildings served by unitary or packaged HVAC equipment listed in Tables 503.2.3(1) through 503.2.3(5), each serving one zone and controlled by a single thermostat in the zone served. It also applies to two-pipe heating systems serving one or more zones, where no cooling system is installed.

This section does not apply to fan systems serving multiple zones, nonunitary or nonpackaged HVAC equipment and systems or hydronic or steam heating and hydronic cooling equipment and distribution systems that provide cooling or cooling and heating which are covered by Section 503.4.

503.3.1 Economizers. Supply air economizers shall be provided on each cooling system as shown in Table 503.3.1(1).

Economizers shall be capable of providing 100-percent outdoor air, even if additional mechanical cooling is required to meet the cooling load of the building. Systems shall provide a means to relieve excess outdoor air during economizer operation to prevent overpressurizing the building. The relief air outlet shall be located to avoid recirculation into the building. Where a single room or space is supplied by multiple air systems, the aggregate capacity of those systems shall be used in applying this requirement.

Exceptions:

1. Where the cooling equipment is covered by the minimum efficiency requirements of Table 503.2.3(1) or 503.2.3(2) and meets or exceeds the minimum cooling efficiency requirement (EER) by the percentages shown in Table 503.3.1(2).

2. Systems with air or evaporatively cooled condensers and which serve spaces with open case refrigeration or that require filtration equipment in order to meet the minimum ventilation require-

ments of Chapter 4 of the *International Mechanical Code.*

TABLE 503.3.1(1)
ECONOMIZER REQUIREMENTS

CLIMATE ZONES	ECONOMIZER REQUIREMENT
1A, 1B, 2A, 7, 8	No requirement
2B, 3A, 3B, 3C, 4A, 4B, 4C, 5A, 5B, 5C, 6A, 6B	Economizers on all cooling systems ≥ 54,000 Btu/h[a]

For SI: 1 British thermal unit per hour = 0.293 W.

a. The total capacity of all systems without economizers shall not exceed 480,000 Btu/h per building, or 20 percent of its air economizer capacity, whichever is greater.

TABLE 503.3.1(2)
EQUIPMENT EFFICIENCY PERFORMANCE
EXCEPTION FOR ECONOMIZERS

CLIMATE ZONES	COOLING EQUIPMENT PERFORMANCE IMPROVEMENT (EER OR IPLV)
2B	10% Efficiency Improvement
3B	15% Efficiency Improvement
4B	20% Efficiency Improvement

503.3.2 Hydronic system controls. Hydronic systems of at least 300,000 Btu/h (87,930 W) design output capacity supplying heated and chilled water to comfort conditioning systems shall include controls that meet the requirements of Section 503.4.3.

503.4 Complex HVAC systems and equipment. (Prescriptive). This section applies to buildings served by HVAC equipment and systems not covered in Section 503.3.

503.4.1 Economizers. Supply air economizers shall be provided on each cooling system according to Table 503.3.1(1). Economizers shall be capable of operating at 100 percent outside air, even if additional mechanical cooling is required to meet the cooling load of the building.

Exceptions:

1. Systems utilizing water economizers that are capable of cooling supply air by direct or indirect evaporation or both and providing 100 percent of the expected system cooling load at outside air temperatures of 50°F (10°C) dry bulb/45°F (7°C) wet bulb and below.

2. Where the cooling equipment is covered by the minimum efficiency requirements of Table 503.2.3(1), 503.2.3(2), or 503.2.3(6) and meets or exceeds the minimum EER by the percentages shown in Table 503.3.1(2)

3. Where the cooling equipment is covered by the minimum efficiency requirements of Table 503.2.3(7) and meets or exceeds the minimum integrated part load value (IPLV) by the percentages shown in Table 503.3.1(2).

503.4.2 Variable air volume (VAV) fan control. Individual VAV fans with motors of 10 horsepower (7.5 kW) or greater shall be:

1. Driven by a mechanical or electrical variable speed drive; or

2. The fan motor shall have controls or devices that will result in fan motor demand of no more than 30 percent of their design wattage at 50 percent of design airflow when static pressure set point equals one-third of the total design static pressure, based on manufacturer's certified fan data.

For systems with direct digital control of individual *zone* boxes reporting to the central control panel, the static pressure set point shall be reset based on the *zone* requiring the most pressure, i.e., the set point is reset lower until one *zone* damper is nearly wide open.

503.4.3 Hydronic systems controls. The heating of fluids that have been previously mechanically cooled and the cooling of fluids that have been previously mechanically heated shall be limited in accordance with Sections 503.4.3.1 through 503.4.3.3. Hydronic heating systems comprised of multiple-packaged boilers and designed to deliver conditioned water or steam into a common distribution system shall include automatic controls capable of sequencing operation of the boilers. Hydronic heating systems comprised of a single boiler and greater than 500,000 Btu/h input design capacity shall include either a multistaged or modulating burner.

503.4.3.1 Three-pipe system. Hydronic systems that use a common return system for both hot water and chilled water are prohibited.

503.4.3.2 Two-pipe changeover system. Systems that use a common distribution system to supply both heated and chilled water shall be designed to allow a dead band between changeover from one mode to the other of at least 15°F (8.3°C) outside air temperatures; be designed to and provided with controls that will allow operation in one mode for at least 4 hours before changing over to the other mode; and be provided with controls that allow heating and cooling supply temperatures at the changeover point to be no more than 30°F (16.7°C) apart.

503.4.3.3 Hydronic (water loop) heat pump systems. Hydronic heat pump systems shall comply with Sections 503.4.3.3.1 through 503.4.3.3.3.

503.4.3.3.1 Temperature dead band. Hydronic heat pumps connected to a common heat pump water loop with central devices for heat rejection and heat addition shall have controls that are capable of providing a heat pump water supply temperature dead band of at least 20°F (11.1°C) between initiation of heat rejection and heat addition by the central devices.

Exception: Where a system loop temperature optimization controller is installed and can determine the most efficient operating temperature based on realtime conditions of demand and capacity, dead bands of less than 20°F (11°C) shall be permitted.

503.4.3.3.2 Heat rejection. Heat rejection equipment shall comply with Sections 503.4.3.3.2.1 and 503.4.3.3.2.2.

Exception: Where it can be demonstrated that a heat pump system will be required to reject heat throughout the year.

503.4.3.3.2.1 Climate Zones 3 and 4. For Climate Zones 3 and 4 as indicated in Figure 301.1 and Table 301.1:

1. If a closed-circuit cooling tower is used directly in the heat pump loop, either an automatic valve shall be installed to bypass all but a minimal flow of water around the tower, or lower leakage positive closure dampers shall be provided.

2. If an open-circuit tower is used directly in the heat pump loop, an automatic valve shall be installed to bypass all heat pump water flow around the tower.

3. If an open- or closed-circuit cooling tower is used in conjunction with a separate heat exchanger to isolate the cooling tower from the heat pump loop, then heat loss shall be controlled by shutting down the circulation pump on the cooling tower loop.

503.4.3.3.2.2 Climate Zones 5 through 8. For climate Zones 5 through 8 as indicated in Figure 301.1 and Table 301.1, if an open- or closed-circuit cooling tower is used, then a separate heat exchanger shall be required to isolate the cooling tower from the heat pump loop, and heat loss shall be controlled by shutting down the circulation pump on the cooling tower loop and providing an automatic valve to stop the flow of fluid.

503.4.3.3.3 Two position valve. Each hydronic heat pump on the hydronic system having a total pump system power exceeding 10 horsepower (hp) (7.5 kW) shall have a two-position valve.

503.4.3.4 Part load controls. Hydronic systems greater than or equal to 300,000 Btu/h (87 930 W) in design output capacity supplying heated or chilled water to comfort conditioning systems shall include controls that have the capability to:

1. Automatically reset the supply-water temperatures using zone-return water temperature, building-return water temperature, or outside air temperature as an indicator of building heating or cooling demand. The temperature shall be capable of being reset by at least 25 percent of the design supply-to-return water temperature difference; or

2. Reduce system pump flow by at least 50 percent of design flow rate utilizing adjustable speed drive(s) on pump(s), or multiple-staged pumps where at least one-half of the total pump horsepower is capable of being automatically turned off or control valves designed to modulate or step down, and

close, as a function of load, or other *approved* means.

503.4.3.5 Pump isolation. Chilled water plants including more than one chiller shall have the capability to reduce flow automatically through the chiller plant when a chiller is shut down. Chillers piped in series for the purpose of increased temperature differential shall be considered as one chiller.

Boiler plants including more than one boiler shall have the capability to reduce flow automatically through the boiler plant when a boiler is shut down.

503.4.4 Heat rejection equipment fan speed control. Each fan powered by a motor of 7.5 hp (5.6 kW) or larger shall have the capability to operate that fan at two-thirds of full speed or less, and shall have controls that automatically change the fan speed to control the leaving fluid temperature or condensing temperature/pressure of the heat rejection device.

Exception: Factory-installed heat rejection devices within HVAC equipment tested and rated in accordance with Tables 503.2.3(6) and 503.2.3(7).

503.4.5 Requirements for complex mechanical systems serving multiple zones. Sections 503.4.5.1 through 503.4.5.4 shall apply to complex mechanical systems serving multiple zones. Supply air systems serving multiple zones shall be VAV systems which, during periods of occupancy, are designed and capable of being controlled to reduce primary air supply to each *zone* to one of the following before reheating, recooling or mixing takes place:

1. Thirty percent of the maximum supply air to each *zone*.

2. Three hundred cfm (142 L/s) or less where the maximum flow rate is less than 10 percent of the total fan system supply airflow rate.

3. The minimum ventilation requirements of Chapter 4 of the *International Mechanical Code*.

Exception: The following define when individual zones or when entire air distribution systems are exempted from the requirement for VAV control:

1. Zones where special pressurization relationships or cross-contamination requirements are such that VAV systems are impractical.

2. Zones or supply air systems where at least 75 percent of the energy for reheating or for providing warm air in mixing systems is provided from a site-recovered or site-solar energy source.

3. Zones where special humidity levels are required to satisfy process needs.

4. Zones with a peak supply air quantity of 300 cfm (142 L/s) or less and where the flow rate is less than 10 percent of the total fan system supply airflow rate.

5. Zones where the volume of air to be reheated, recooled or mixed is no greater than the volume of outside air required to meet the minimum ventilation requirements of Chapter 4 of the *International Mechanical Code*.

6. Zones or supply air systems with thermostatic and humidistatic controls capable of operating in sequence the supply of heating and cooling energy to the *zone*(s) and which are capable of preventing reheating, recooling, mixing or simultaneous supply of air that has been previously cooled, either mechanically or through the use of economizer systems, and air that has been previously mechanically heated.

503.4.5.1 Single duct variable air volume (VAV) systems, terminal devices. Single duct VAV systems shall use terminal devices capable of reducing the supply of primary supply air before reheating or recooling takes place.

503.4.5.2 Dual duct and mixing VAV systems, terminal devices. Systems that have one warm air duct and one cool air duct shall use terminal devices which are capable of reducing the flow from one duct to a minimum before mixing of air from the other duct takes place.

503.4.5.3 Single fan dual duct and mixing VAV systems, economizers. Individual dual duct or mixing heating and cooling systems with a single fan and with total capacities greater than 90,000 Btu/h [(26 375 W) 7.5 tons] shall not be equipped with air economizers.

503.4.5.4 Supply-air temperature reset controls. Multiple *zone* HVAC systems shall include controls that automatically reset the supply-air temperature in response to representative building loads, or to outdoor air temperature. The controls shall be capable of resetting the supply air temperature at least 25 percent of the difference between the design supply-air temperature and the design room air temperature.

Exceptions:

1. Systems that prevent reheating, recooling or mixing of heated and cooled supply air.

2. Seventy five percent of the energy for reheating is from site-recovered or site solar energy sources.

3. Zones with peak supply air quantities of 300 cfm (142 L/s) or less.

503.4.6 Heat recovery for service water heating. Condenser heat recovery shall be installed for heating or reheating of service hot water provided the facility operates 24 hours a day, the total installed heat capacity of water-cooled systems exceeds 6,000,000 Btu/hr of heat rejection, and the design service water heating load exceeds 1,000,000 Btu/h.

The required heat recovery system shall have the capacity to provide the smaller of:

1. Sixty percent of the peak heat rejection load at design conditions; or

2. The preheating required to raise the peak service hot water draw to 85°F (29°C).

Exceptions:

1. Facilities that employ condenser heat recovery for space heating or reheat purposes with a heat recovery design exceeding 30 percent of the peak water-cooled condenser load at design conditions.

2. Facilities that provide 60 percent of their service water heating from site solar or site recovered energy or from other sources.

503.4.7 Hot gas bypass limitation. Cooling systems shall not use hot gas bypass or other evaporator pressure control systems unless the system is designed with multiple steps of unloading or continuous capacity modulation. The capacity of the hot gas bypass shall be limited as indicated in Table 503.4.7.

Exception: Unitary packaged systems with cooling capacities not greater than 90,000 Btu/h (26 379 W).

TABLE 503.4.7
MAXIMUM HOT GAS BYPASS CAPACITY

RATED CAPACITY	MAXIMUM HOT GAS BYPASS CAPACITY (% of total capacity)
≤ 240,000 Btu/h	50%
> 240,000 Btu/h	25%

For SI: 1 Btu/h = 0.29 watts.

SECTION 504
SERVICE WATER HEATING
(Mandatory)

504.1 General. This section covers the minimum efficiency of, and controls for, service water-heating equipment and insulation of service hot water piping.

504.2 Service water-heating equipment performance efficiency. Water-heating equipment and hot water storage tanks shall meet the requirements of Table 504.2. The efficiency shall be verified through data furnished by the manufacturer or through certification under an *approved* certification program.

504.3 Temperature controls. Service water-heating equipment shall be provided with controls to allow a setpoint of 110°F (43°C) for equipment serving dwelling units and 90°F (32°C) for equipment serving other occupancies. The outlet temperature of lavatories in public facility rest rooms shall be limited to 110°F (43°C).

504.4 Heat traps. Water-heating equipment not supplied with integral heat traps and serving noncirculating systems shall be provided with heat traps on the supply and discharge piping associated with the equipment.

504.5 Pipe insulation. For automatic-circulating hot water systems, piping shall be insulated with 1 inch (25 mm) of insulation having a conductivity not exceeding 0.27 Btu per inch/h × ft² × °F (1.53 W per 25 mm/m² × K). The first 8 feet (2438 mm) of piping in noncirculating systems served by equipment without integral heat traps shall be insulated with 0.5 inch (12.7 mm) of material having a conductivity not exceeding 0.27 Btu per inch/h × ft² × °F (1.53 W per 25 mm/m² × K).

504.6 Hot water system controls. Automatic-circulating hot water system pumps or heat trace shall be arranged to be conveniently turned off automatically or manually when the hot water system is not in operation.

504.7 Pools. Pools shall be provided with energy conserving measures in accordance with Sections 504.7.1 through 504.7.3.

504.7.1 Pool heaters. All pool heaters shall be equipped with a readily *accessible* on-off switch to allow shutting off the heater without adjusting the thermostat setting. Pool heaters fired by natural gas or LPG shall not have continuously burning pilot lights.

504.7.2 Time switches. Time switches that can automatically turn off and on heaters and pumps according to a preset schedule shall be installed on swimming pool heaters and pumps.

Exceptions:

1. Where public health standards require 24-hour pump operation.

2. Where pumps are required to operate solar-and waste-heat-recovery pool heating systems.

504.7.3 Pool covers. Heated pools shall be equipped with a vapor retardant pool cover on or at the water surface. Pools heated to more than 90°F (32°C) shall have a pool cover with a minimum insulation value of R-12.

Exception: Pools deriving over 60 percent of the energy for heating from site-recovered energy or solar energy source.

SECTION 505
ELECTRICAL POWER AND LIGHTING SYSTEMS
(Prescriptive)

505.1 General (Prescriptive). This section covers lighting system controls, the connection of ballasts, the maximum lighting power for interior applications and minimum acceptable lighting equipment for exterior applications.

Exception: Lighting within dwelling units where 50 percent or more of the permanently installed interior light fixtures are fitted with high-efficacy lamps.

505.2 Lighting controls (Prescriptive). Lighting systems shall be provided with controls as required in Sections 505.2.1, 505.2.2, 505.2.3 and 505.2.4.

TABLE 504.2
MINIMUM PERFORMANCE OF WATER-HEATING EQUIPMENT

EQUIPMENT TYPE	SIZE CATEGORY (Input)	SUBCATEGORY OR RATING CONDITION	PERFORMANCE REQUIRED[a, b]	TEST PROCEDURE
Water heaters, Electric	≤ 12 kW	Resistance	$0.97 - 0.00132V$, EF	DOE 10 CFR Part 430
	> 12 kW	Resistance	$1.73V + 155$ SL, Btu/h	ANSI Z21.10.3
	≤ 24 amps and ≤ 250 volts	Heat pump	$0.93 - 0.00132V$, EF	DOE 10 CFR Part 430
Storage water heaters, Gas	≤ 75,000 Btu/h	≥ 20 gal	$0.67 - 0.0019V$, EF	DOE 10 CFR Part 430
	> 75,000 Btu/h and ≤ 155,000 Btu/h	< 4,000 Btu/h/gal	$80\% \, E_t$ $\left(Q/800 + 110\sqrt{V}\right)$ SL, Btu/h	ANSI Z21.10.3
	> 155,000 Btu/h	< 4,000 Btu/h/gal	$80\% \, E_t$ $\left(Q/800 + 110\sqrt{V}\right)$ SL, Btu/h	
Instantaneous water heaters, Gas	> 50,000 Btu/h and < 200,000 Btu/h[c]	≥ 4,000 (Btu/h)/gal and < 2 gal	$0.62 - 0.0019V$, EF	DOE 10 CFR Part 430
	≥ 200,000 Btu/h	≥ 4,000 Btu/h/gal and < 10 gal	$80\% \, E_t$	ANSI Z21.10.3
	≥ 200,000 Btu/h	≥ 4,000 Btu/h/gal and ≥ 10 gal	$80\% \, E_t$ $\left(Q/800 + 110\sqrt{V}\right)$ SL, Btu/h	
Storage water heaters, Oil	≤ 105,000 Btu/h	≥ 20 gal	$0.59 - 0.0019V$, EF	DOE 10 CFR Part 430
	> 105,000 Btu/h	< 4,000 Btu/h/gal	$78\% \, E_t$ $\left(Q/800 + 110\sqrt{V}\right)$ SL, Btu/h	ANSI Z21.10.3
Instantaneous water heaters, Oil	≤ 210,000 Btu/h	≥ 4,000 Btu/h/gal and < 2 gal	$0.59 - 0.0019V$, EF	DOE 10 CFR Part 430
	> 210,000 Btu/h	≥ 4,000 Btu/h/gal and < 10 gal	$80\% \, E_t$	ANSI Z21.10.3
	> 210,000 Btu/h	≥ 4,000 Btu/h/gal and ≥ 10 gal	$78\% \, E_t$ $\left(Q/800 + 110\sqrt{V}\right)$ SL, Btu/h	
Hot water supply boilers, Gas and Oil	≥ 300,000 Btu/h and <12,500,000 Btu/h	≥ 4,000 Btu/h/gal and < 10 gal	$80\% \, E_t$	
Hot water supply boilers, Gas	≥ 300,000 Btu/h and <12,500,000 Btu/h	≥ 4,000 Btu/h/gal and ≥ 10 gal	$80\% \, E_t$ $\left(Q/800 + 110\sqrt{V}\right)$ SL, Btu/h	ANSI Z21.10.3
Hot water supply boilers, Oil	> 300,000 Btu/h and <12,500,000 Btu/h	> 4,000 Btu/h/gal and > 10 gal	$78\% \, E_t$ $\left(Q/800 + 110\sqrt{V}\right)$ SL, Btu/h	
Pool heaters, Gas and Oil	All	—	$78\% \, E_t$	ASHRAE 146
Heat pump pool heaters	All	—	4.0 COP	AHRI 1160
Unfired storage tanks	All	—	Minimum insulation requirement R-12.5 $(h \cdot ft^2 \cdot °F)/Btu$	(none)

For SI: °C = [(°F) - 32]/1.8, 1 British thermal unit per hour = 0.2931 W, 1 gallon = 3.785 L, 1 British thermal unit per hour per gallon = 0.078 W/L.

a. Energy factor (EF) and thermal efficiency (E_t) are minimum requirements. In the EF equation, V is the rated volume in gallons.

b. Standby loss (SL) is the maximum Btu/h based on a nominal 70°F temperature difference between stored water and ambient requirements. In the SL equation, Q is the nameplate input rate in Btu/h. In the SL equation for electric water heaters, V is the rated volume in gallons. In the SL equation for oil and gas water heaters and boilers, V is the rated volume in gallons.

c. Instantaneous water heaters with input rates below 200,000 Btu/h must comply with these requirements if the water heater is designed to heat water to temperatures 180°F or higher.

505.2.1 Interior lighting controls. Each area enclosed by walls or floor-to-ceiling partitions shall have at least one manual control for the lighting serving that area. The required controls shall be located within the area served by the controls or be a remote switch that identifies the lights served and indicates their status.

Exceptions:

1. Areas designated as security or emergency areas that must be continuously lighted.

2. Lighting in stairways or corridors that are elements of the means of egress.

505.2.2 Additional controls. Each area that is required to have a manual control shall have additional controls that meet the requirements of Sections 505.2.2.1 and 505.2.2.2.

505.2.2.1 Light reduction controls. Each area that is required to have a manual control shall also allow the occupant to reduce the connected lighting load in a reasonably uniform illumination pattern by at least 50 percent. Lighting reduction shall be achieved by one of the following or other *approved* method:

1. Controlling all lamps or luminaires;

2. Dual switching of alternate rows of luminaires, alternate luminaires or alternate lamps;

3. Switching the middle lamp luminaires independently of the outer lamps; or

4. Switching each luminaire or each lamp.

Exceptions:

1. Areas that have only one luminaire.

2. Areas that are controlled by an occupant-sensing device.

3. Corridors, storerooms, restrooms or public lobbies.

4. *Sleeping unit* (see Section 505.2.3).

5. Spaces that use less than 0.6 watts per square foot (6.5 W/m²).

505.2.2.2 Automatic lighting shutoff. Buildings larger than 5,000 square feet (465 m²) shall be equipped with an automatic control device to shut off lighting in those areas. This automatic control device shall function on either:

1. A scheduled basis, using time-of-day, with an independent program schedule that controls the interior lighting in areas that do not exceed 25,000 square feet (2323 m²) and are not more than one floor; or

2. An occupant sensor that shall turn lighting off within 30 minutes of an occupant leaving a space; or

3. A signal from another control or alarm system that indicates the area is unoccupied.

Exception: The following shall not require an automatic control device:

1. *Sleeping unit* (see Section 505.2.3).

2. Lighting in spaces where patient care is directly provided.

3. Spaces where an automatic shutoff would endanger occupant safety or security.

505.2.2.2.1 Occupant override. Where an automatic time switch control device is installed to comply with Section 505.2.2.2, Item 1, it shall incorporate an override switching device that:

1. Is readily *accessible*.

2. Is located so that a person using the device can see the lights or the area controlled by that switch, or so that the area being lit is annunciated.

3. Is manually operated.

4. Allows the lighting to remain on for no more than 2 hours when an override is initiated.

5. Controls an area not exceeding 5,000 square feet (465 m²).

Exceptions:

1. In malls and arcades, auditoriums, single-tenant retail spaces, industrial facilities and arenas, where captive-key override is utilized, override time shall be permitted to exceed 2 hours.

2. In malls and arcades, auditoriums, single-tenant retail spaces, industrial facilities and arenas, the area controlled shall not exceed 20,000 square feet (1860 m²).

505.2.2.2.2 Holiday scheduling. If an automatic time switch control device is installed in accordance with Section 505.2.2.2, Item 1, it shall incorporate an automatic holiday scheduling feature that turns off all loads for at least 24 hours, then resumes the normally scheduled operation.

Exception: Retail stores and associated malls, restaurants, grocery stores, places of religious worship and theaters.

505.2.2.3 Daylight zone control. Daylight zones, as defined by this code, shall be provided with individual controls that control the lights independent of general area lighting. Contiguous daylight zones adjacent to vertical fenestration are allowed to be controlled by a single controlling device provided that they do not include zones facing more than two adjacent cardinal orientations (i.e., north, east, south, west). Daylight zones under skylights more than 15 feet (4572 mm) from the perimeter shall be controlled separately from daylight zones adjacent to vertical fenestration.

Exception: Daylight spaces enclosed by walls or ceiling height partitions and containing two or fewer light fixtures are not required to have a separate switch for general area lighting.

505.2.3 Sleeping unit controls. *Sleeping units* in hotels, motels, boarding houses or similar buildings shall have at least one master switch at the main entry door that controls all permanently wired luminaires and switched receptacles, except those in the bathroom(s). Suites shall have a control meeting these requirements at the entry to each room or at the primary entry to the suite.

505.2.4 Exterior lighting controls. Lighting not designated for dusk-to-dawn operation shall be controlled by either a combination of a photosensor and a time switch, or an astronomical time switch. Lighting designated for dusk-to-dawn operation shall be controlled by an astronomical time switch or photosensor. All time switches shall be capable of retaining programming and the time setting during loss of power for a period of at least 10 hours.

505.3 Tandem wiring (Prescriptive). The following luminaires located within the same area shall be tandem wired:

1. Fluorescent luminaires equipped with one, three or odd-numbered lamp configurations, that are recess-mounted within 10 feet (3048 mm) center-to-center of each other.

2. Fluorescent luminaires equipped with one, three or any odd-numbered lamp configuration, that are pendant- or surface-mounted within 1 foot (305 mm) edge- to-edge of each other.

Exceptions:

1. Where electronic high-frequency ballasts are used.

2. Luminaires on emergency circuits.

3. Luminaires with no available pair in the same area.

505.4 Exit signs (Prescriptive). Internally illuminated exit signs shall not exceed 5 watts per side.

505.5 Interior lighting power requirements (Prescriptive). A building complies with this section if its total connected lighting power calculated under Section 505.5.1 is no greater than the interior lighting power calculated under Section 505.5.2.

505.5.1 Total connected interior lighting power. The total connected interior lighting power (watts) shall be the sum of the watts of all interior lighting equipment as determined in accordance with Sections 505.5.1.1 through 505.5.1.4.

Exceptions:

1. The connected power associated with the following lighting equipment is not included in calculating total connected lighting power.

   1.1. Professional sports arena playing field lighting.

   1.2. *Sleeping unit* lighting in hotels, motels, boarding houses or similar buildings.

   1.3. Emergency lighting automatically off during normal building operation.

   1.4. Lighting in spaces specifically designed for use by occupants with special lighting needs including the visually impaired visual impairment and other medical and age-related issues.

   1.5. Lighting in interior spaces that have been specifically designated as a registered interior historic landmark.

   1.6. Casino gaming areas.

2. Lighting equipment used for the following shall be exempt provided that it is in addition to general lighting and is controlled by an independent control device:

   2.1. Task lighting for medical and dental purposes.

   2.2. Display lighting for exhibits in galleries, museums and monuments.

3. Lighting for theatrical purposes, including performance, stage, film production and video production.

4. Lighting for photographic processes.

5. Lighting integral to equipment or instrumentation and is installed by the manufacturer.

6. Task lighting for plant growth or maintenance.

7. Advertising signage or directional signage.

8. In restaurant buildings and areas, lighting for food warming or integral to food preparation equipment.

9. Lighting equipment that is for sale.

10. Lighting demonstration equipment in lighting education facilities.

11. Lighting *approved* because of safety or emergency considerations, inclusive of exit lights.

12. Lighting integral to both open and glass-enclosed refrigerator and freezer cases.

13. Lighting in retail display windows, provided the display area is enclosed by ceiling-height partitions.

14. Furniture mounted supplemental task lighting that is controlled by automatic shutoff.

505.5.1.1 Screw lamp holders. The wattage shall be the maximum *labeled* wattage of the luminaire.

505.5.1.2 Low-voltage lighting. The wattage shall be the specified wattage of the transformer supplying the system.

505.5.1.3 Other luminaires. The wattage of all other lighting equipment shall be the wattage of the lighting equipment verified through data furnished by the manufacturer or other *approved* sources.

505.5.1.4 Line-voltage lighting track and plug-in busway. The wattage shall be:

1. The specified wattage of the luminaires included in the system with a minimum of 30 W/lin ft. (98 W/lin. m);

2. The wattage limit of the system's circuit breaker; or

3. The wattage limit of other permanent current limiting device(s) on the system.

505.5.2 Interior lighting power. The total interior lighting power (watts) is the sum of all interior lighting powers for all areas in the building covered in this permit. The interior lighting power is the floor area for each building area type listed in Table 505.5.2 times the value from Table 505.5.2 for that area. For the purposes of this method, an "area" shall be defined as all contiguous spaces that accommodate or are associated with a single building area type as *listed* in Table 505.5.2. When this method is used to calculate the total interior lighting power for an entire building, each building area type shall be treated as a separate area.

TABLE 505.5.2
INTERIOR LIGHTING POWER ALLOWANCES

LIGHTING POWER DENSITY	
Building Area Type[a]	(W/ft²)
Automotive Facility	0.9
Convention Center	1.2
Court House	1.2
Dining: Bar Lounge/Leisure	1.3
Dining: Cafeteria/Fast Food	1.4
Dining: Family	1.6
Dormitory	1.0
Exercise Center	1.0
Gymnasium	1.1
Healthcare—clinic	1.0
Hospital	1.2
Hotel	1.0
Library	1.3
Manufacturing Facility	1.3
Motel	1.0
Motion Picture Theater	1.2
Multifamily	0.7
Museum	1.1
Office	1.0
Parking Garage	0.3

(continued)

TABLE 505.5.2—continued
INTERIOR LIGHTING POWER ALLOWANCES

LIGHTING POWER DENSITY	
Building Area Type[a]	(W/ft²)
Penitentiary	1.0
Performing Arts Theater	1.6
Police/Fire Station	1.0
Post Office	1.1
Religious Building	1.3
Retail[b]	1.5
School/University	1.2
Sports Arena	1.1
Town Hall	1.1
Transportation	1.0
Warehouse	0.8
Workshop	1.4

For SI: 1 foot = 304.8 mm, 1 watt per square foot = W/0.0929 m².

a. In cases where both a general building area type and a more specific building area type are listed, the more specific building area type shall apply.

b. Where lighting equipment is specified to be installed to highlight specific merchandise in addition to lighting equipment specified for general lighting and is switched or dimmed on circuits different from the circuits for general lighting, the smaller of the actual wattage of the lighting equipment installed specifically for merchandise, or additional lighting power as determined below shall be added to the interior lighting power determined in accordance with this line item.

Calculate the additional lighting power as follows:

Additional Interior Lighting Power Allowance = 1000 watts + (Retail Area 1 × 0.6 W/ft²) + (Retail Area 2 × 0.6 W/ft²) + (Retail Area 3 × 1.4 W/ft²) + (Retail Area 4 × 2.5 W/ft²).

where:

Retail Area 1 = The floor area for all products not listed in Retail Area 2, 3 or 4.

Retail Area 2 = The floor area used for the sale of vehicles, sporting goods and small electronics.

Retail Area 3 = The floor area used for the sale of furniture, clothing, cosmetics and artwork.

Retail Area 4 = The floor area used for the sale of jewelry, crystal and china.

Exception: Other merchandise categories are permitted to be included in Retail Areas 2 through 4 above, provided that justification documenting the need for additional lighting power based on visual inspection, contrast, or other critical display is *approved* by the authority having jurisdiction.

505.6 Exterior lighting. (Prescriptive). When the power for exterior lighting is supplied through the energy service to the building, all exterior lighting, other than low-voltage landscape lighting, shall comply with Sections 505.6.1 and 505.6.2.

Exception: Where *approved* because of historical, safety, signage or emergency considerations.

505.6.1 Exterior building grounds lighting. All exterior building grounds luminaires that operate at greater than 100 watts shall contain lamps having a minimum efficacy of 60 lumens per watt unless the luminaire is controlled by a motion sensor or qualifies for one of the exceptions under Section 505.6.2.

505.6.2 Exterior building lighting power. The total exterior lighting power allowance for all exterior building applications is the sum of the base site allowance plus the individual allowances for areas that are to be illuminated and are permitted in Table 505.6.2(2) for the applicable lighting *zone*. Tradeoffs are allowed only among exterior lighting applications listed in Table 505.6.2(2), Tradable Surfaces section. The lighting zone for the building exterior is determined from Table 505.6.2(1) unless otherwise specified by the local jurisdiction. Exterior lighting for all applications (except those included in the exceptions to Section 505.6.2) shall comply with the requirements of Section 505.6.1.

> **Exceptions:** Lighting used for the following exterior applications is exempt when equipped with a control device independent of the control of the nonexempt lighting:
>
> 1. Specialized signal, directional and marker lighting associated with transportation;
>
> 2. Advertising signage or directional signage;
>
> 3. Integral to equipment or instrumentation and is installed by its manufacturer;
>
> 4. Theatrical purposes, including performance, stage, film production and video production;
>
> 5. Athletic playing areas;
>
> 6. Temporary lighting;
>
> 7. Industrial production, material handling, transportation sites and associated storage areas;
>
> 8. Theme elements in theme/amusement parks; and
>
> 9. Used to highlight features of public monuments and registered historic landmark structures or buildings.

TABLE 505.6.2(1)
EXTERIOR LIGHTING ZONES

LIGHTING ZONE	DESCRIPTION
1	Developed areas of national parks, state parks, forest land, and rural areas
2	Areas predominantly consisting of residential zoning, neighborhood business districts, light industrial with limited nighttime use and residential mixed use areas
3	All other areas
4	High-activity commercial districts in major metropolitan areas as designated by the local land use planning authority

505.7 Electrical energy consumption. (Prescriptive). In buildings having individual dwelling units, provisions shall be made to determine the electrical energy consumed by each tenant by separately metering individual dwelling units.

SECTION 506
TOTAL BUILDING PERFORMANCE

506.1 Scope. This section establishes criteria for compliance using total building performance. The following systems and loads shall be included in determining the total building performance: heating systems, cooling systems, service water heating, fan systems, lighting power, receptacle loads and process loads.

506.2 Mandatory requirements. Compliance with this section requires that the criteria of Sections 502.4, 503.2, 504 and 505 be met.

506.3 Performance-based compliance. Compliance based on total building performance requires that a proposed building (*proposed design*) be shown to have an annual energy cost that is less than or equal to the annual energy cost of the *standard reference design*. Energy prices shall be taken from a source *approved* by the *code official*, such as the Department of Energy, Energy Information Administration's *State Energy Price and Expenditure Report*. *Code officials* shall be permitted to require time-of-use pricing in energy cost calculations. Nondepletable energy collected off site shall be treated and priced the same as purchased energy. Energy from nondepletable energy sources collected on site shall be omitted from the annual energy cost of the *proposed design*.

> **Exception:** Jurisdictions that require site energy (1 kWh = 3413 Btu) rather than energy cost as the metric of comparison.

506.4 Documentation. Documentation verifying that the methods and accuracy of compliance software tools conform to the provisions of this section shall be provided to the *code official*.

506.4.1 Compliance report. Compliance software tools shall generate a report that documents that the *proposed design* has annual energy costs less than or equal to the annual energy costs of the *standard reference design*. The compliance documentation shall include the following information:

1. Address of the building;

2. An inspection checklist documenting the building component characteristics of the *proposed design* as *listed* in Table 506.5.1(1). The inspection checklist shall show the estimated annual energy cost for both the *standard reference design* and the *proposed design*;

3. Name of individual completing the compliance report; and

4. Name and version of the compliance software tool.

TABLE 505.6.2(2)
INDIVIDUAL LIGHTING POWER ALLOWANCES FOR BUILDING EXTERIORS

		Zone 1	Zone 2	Zone 3	Zone 4
Base Site Allowance (Base allowance may be used in tradable or nontradable surfaces.)		500 W	600 W	750 W	1300 W
Tradable Surfaces (Lighting power densities for uncovered parking areas, building grounds, building entrances and exits, canopies and overhangs and outdoor sales areas may be traded.)	**Uncovered Parking Areas**				
	Parking areas and drives	0.04 W/ft²	0.06 W/ft²	0.10 W/ft²	0.13 W/ft²
	Building Grounds				
	Walkways less than 10 feet wide	0.7 W/linear foot	0.7 W/linear foot	0.8 W/linear foot	1.0 W/linear foot
	Walkways 10 feet wide or greater, plaza areas special feature areas	0.14 W/ft²	0.14 W/ft²	0.16 W/ft²	0.2 W/ft²
	Stairways	0.75 W/ft²	1.0 W/ft²	1.0 W/ft²	1.0 W/ft²
	Pedestrian tunnels	0.15 W/ft²	0.15 W/ft²	0.2 W/ft²	0.3 W/ft²
	Building Entrances and Exits				
	Main entries	20 W/linear foot of door width	20 W/linear foot of door width	30 W/linear foot of door width	30 W/linear foot of door width
	Other doors	20 W/linear foot of door width	20 W/linear foot of door width	20 W/linear foot of door width	20 W/linear foot of door width
	Entry canopies	0.25 W/ft²	0.25 W/ft²	0.4 W/ft²	0.4 W/ft²
	Sales Canopies				
	Free-standing and attached	0.6 W/ft²	0.6 W/ft²	0.8 W/ft²	1.0 W/ft²
	Outdoor Sales				
	Open areas (including vehicle sales lots)	0.25 W/ft²	0.25 W/ft²	0.5 W/ft²	0.7 W/ft²
	Street frontage for vehicle sales lots in addition to "open area" allowance	No allowance	10 W/linear foot	10 W/linear foot	30 W/linear foot
Nontradable Surfaces (Lighting power density calculations for the following applications can be used only for the specific application and cannot be traded between surfaces or with other exterior lighting. The following allowances are in addition to any allowance otherwise permitted in the "Tradable Surfaces" section of this table.)	Building facades	No allowance	0.1 W/ft² for each illuminated wall or surface or 2.5 W/linear foot for each illuminated wall or surface length	0.15 W/ft² for each illuminated wall or surface or 3.75 W/linear foot for each illuminated wall or surface length	0.2 W/ft² for each illuminated wall or surface or 5.0 W/linear foot for each illuminated wall or surface length
	Automated teller machines and night depositories	270 W per location plus 90 W per additional ATM per location	270 W per location plus 90 W per additional ATM per location	270 W per location plus 90 W per additional ATM per location	270 W per location plus 90 W per additional ATM per location
	Entrances and gatehouse inspection stations at guarded facilities	0.75 W/ft² of covered and uncovered area	0.75 W/ft² of covered and uncovered area	0.75 W/ft² of covered and uncovered area	0.75 W/ft² of covered and uncovered area
	Loading areas for law enforcement, fire, ambulance and other emergency service vehicles	0.5 W/ft² of covered and uncovered area	0.5 W/ft² of covered and uncovered area	0.5 W/ft² of covered and uncovered area	0.5 W/ft² of covered and uncovered area
	Drive-up windows/doors	400 W per drive-through	400 W per drive-through	400 W per drive-through	400 W per drive-through
	Parking near 24-hour retail entrances	800 W per main entry	800 W per main entry	800 W per main entry	800 W per main entry

For SI: 1 foot = 304.8 mm, 1 watt per square foot = W/0.0929 m².

506.4.2 Additional documentation. The *code official* shall be permitted to require the following documents:

1. Documentation of the building component characteristics of the *standard reference design*;

2. Thermal zoning diagrams consisting of floor plans showing the thermal zoning scheme for *standard reference design* and *proposed design*.

3. Input and output report(s) from the energy analysis simulation program containing the complete input and output files, as applicable. The output file shall include energy use totals and energy use by energy source and end-use served, total hours that space conditioning loads are not met and any errors or warning messages generated by the simulation tool as applicable;

4. An explanation of any error or warning messages appearing in the simulation tool output; and

5. A certification signed by the builder providing the building component characteristics of the *proposed design* as given in Table 506.5.1(1).

506.5 Calculation procedure. Except as specified by this section, the *standard reference design* and *proposed design* shall be configured and analyzed using identical methods and techniques.

506.5.1 Building specifications. The *standard reference design* and *proposed design* shall be configured and analyzed as specified by Table 506.5.1(1). Table 506.5.1(1) shall include by reference all notes contained in Table 502.2(1).

506.5.2 Thermal blocks. The *standard reference design* and *proposed design* shall be analyzed using identical thermal blocks as required in Section 506.5.2.1, 506.5.2.2 or 506.5.2.3.

506.5.2.1 HVAC zones designed. Where HVAC zones are defined on HVAC design drawings, each HVAC *zone* shall be modeled as a separate thermal block.

Exception: Different HVAC zones shall be allowed to be combined to create a single thermal block or identical thermal blocks to which multipliers are applied provided:

1. The space use classification is the same throughout the thermal block.

2. All HVAC zones in the thermal block that are adjacent to glazed exterior walls face the same orientation or their orientations are within 45 degrees (0.79 rad) of each other.

3. All of the zones are served by the same HVAC system or by the same kind of HVAC system.

506.5.2.2 HVAC zones not designed. Where HVAC zones have not yet been designed, thermal blocks shall be defined based on similar internal load densities, occupancy, lighting, thermal and temperature schedules, and in combination with the following guidelines:

1. Separate thermal blocks shall be assumed for interior and perimeter spaces. Interior spaces shall be those located more than 15 feet (4572 mm) from an exterior wall. Perimeter spaces shall be those located closer than 15 feet (4572 mm) from an *exterior wall*.

2. Separate thermal blocks shall be assumed for spaces adjacent to glazed exterior walls: a separate *zone* shall be provided for each orientation, except orientations that differ by no more than 45 degrees (0.79 rad) shall be permitted to be considered to be the same orientation. Each *zone* shall include floor area that is 15 feet (4572 mm) or less from a glazed perimeter wall, except that floor area within 15 feet (4572 mm) of glazed perimeter walls having more than one orientation shall be divided proportionately between zones.

3. Separate thermal blocks shall be assumed for spaces having floors that are in contact with the ground or exposed to ambient conditions from zones that do not share these features.

4. Separate thermal blocks shall be assumed for spaces having exterior ceiling or roof assemblies from zones that do not share these features.

506.5.2.3 Multifamily residential buildings. Residential spaces shall be modeled using one thermal block per space except that those facing the same orientations are permitted to be combined into one thermal block. Corner units and units with roof or floor loads shall only be combined with units sharing these features.

TABLE 506.5.1(1)
SPECIFICATIONS FOR THE STANDARD REFERENCE AND PROPOSED DESIGNS

BUILDING COMPONENT CHARACTERISTICS	STANDARD REFERENCE DESIGN	PROPOSED DESIGN
Space use classification	Same as proposed	The space use classification shall be chosen in accordance with Table 505.5.2 for all areas of the building covered by this permit. Where the space use classification for a building is not known, the building shall be categorized as an office building.
Roofs	Type: Insulation entirely above deck Gross area: same as proposed U-factor: from Table 502.1.2 Solar absorptance: 0.75 Emittance: 0.90	As proposed As proposed As proposed As proposed As proposed
Walls, above-grade	Type: Mass wall if proposed wall is mass; otherwise steel-framed wall Gross area: same as proposed U-factor: from Table 502.1.2 Solar absorptance: 0.75 Emittance: 0.90	As proposed As proposed As proposed As proposed As proposed
Walls, below-grade	Type: Mass wall Gross area: same as proposed U-Factor: from Table 502.1.2 with insulation layer on interior side of walls	As proposed As proposed As proposed
Floors, above-grade	Type: joist/framed floor Gross area: same as proposed U-factor: from Table 502.1.2	As proposed As proposed As proposed
Floors, slab-on-grade	Type: Unheated F-factor: from Table 502.1.2	As proposed As proposed
Doors	Type: Swinging Area: Same as proposed U-factor: from Table 502.2(1)	As proposed As proposed As proposed
Glazing	Area: (a) The proposed glazing area; where the proposed glazing area is less than 40 percent of above-grade wall area. (b) 40 percent of above-grade wall area; where the proposed glazing area is 40 percent or more of the above-grade wall area. U-factor: from Table 502.3 SHGC: from Table 502.3 except that for climates with no requirement (NR) SHGC = 0.40 shall be used External shading and PF: None	As proposed As proposed As proposed As proposed
Skylights	Area: (a) The proposed skylight area; where the proposed skylight area is less than 3 percent of gross area of roof assembly. (b) 3 percent of gross area of roof assembly; where the proposed skylight area is 3 percent or more of gross area of roof assembly. U-factor: from Table 502.3 SHGC: from Table 502.3 except that for climates with no requirement (NR) SHGC = 0.40 shall be used.	As proposed As proposed As proposed
Lighting, interior	The interior lighting power shall be determined in accordance with Table 505.5.2. Where the occupancy of the building is not known, the lighting power density shall be 1.0 Watt per square foot (10.73 W/m^2) based on the categorization of buildings with unknown space classification as offices.	As proposed
Lighting, exterior	The lighting power shall be determined in accordance with Table 505.6.2(2). Areas and dimensions of tradable and nontradable surfaces shall be the same as proposed.	As proposed

(continued)

TABLE 506.5.1(1)—continued
SPECIFICATIONS FOR THE STANDARD REFERENCE AND PROPOSED DESIGNS

BUILDING COMPONENT CHARACTERISTICS	STANDARD REFERENCE DESIGN	PROPOSED DESIGN
Internal gains	Same as proposed	Receptacle, motor and process loads shall be modeled and estimated based on the space use classification. All end-use load components within and associated with the building shall be modeled to include, but not be limited to, the following: exhaust fans, parking garage ventilation fans, exterior building lighting, swimming pool heaters and pumps, elevators, escalators, refrigeration equipment and cooking equipment.
Schedules	Same as proposed	Operating schedules shall include hourly profiles for daily operation and shall account for variations between weekdays, weekends, holidays and any seasonal operation. Schedules shall model the time-dependent variations in occupancy, illumination, receptacle loads, thermostat settings, mechanical ventilation, HVAC equipment availability, service hot water usage and any process loads. The schedules shall be typical of the proposed building type as determined by the designer and approved by the jurisdiction.
Mechanical ventilation	Same as proposed	As proposed, in accordance with Section 503.2.5.
Heating systems	Fuel type: same as proposed design Equipment type[a]: from Tables 506.5.1(2) and 506.5.1(3) Efficiency: from Tables 503.2.3(4) and 503.2.3(5) Capacity[b]: sized proportionally to the capacities in the proposed design based on sizing runs, and shall be established such that no smaller number of unmet heating load hours and no larger heating capacity safety factors are provided than in the proposed design.	As proposed As proposed As proposed As proposed
Cooling systems	Fuel type: same as proposed design Equipment type[c]: from Tables 506.5.1(2) and 506.5.1(3) Efficiency: from Tables 503.2.3(1), 503.2.3(2) and 503.2.3(3) Capacity[b]: sized proportionally to the capacities in the proposed design based on sizing runs, and shall be established such that no smaller number of unmet cooling load hours and no larger cooling capacity safety factors are provided than in the proposed design. Economizer[d]: same as proposed, in accordance with Section 503.4.1.	As proposed As proposed As proposed As proposed As proposed
Service water heating	Fuel type: same as proposed Efficiency: from Table 504.2 Capacity: same as proposed Where no service water hot water system exists or is specified in the proposed design, no service hot water heating shall be modeled.	As proposed As proposed As proposed

a. Where no heating system exists or has been specified, the heating system shall be modeled as fossil fuel. The system characteristics shall be identical in both the standard reference design and proposed design.

b. The ratio between the capacities used in the annual simulations and the capacities determined by sizing runs shall be the same for both the standard reference design and proposed design.

c. Where no cooling system exists or no cooling system has been specified, the cooling system shall be modeled as an air-cooled single-zone system, one unit per thermal zone. The system characteristics shall be identical in both the standard reference design and proposed design.

d. If an economizer is required in accordance with Table 503.3.1 (1), and if no economizer exists or is specified in the proposed design, then a supply air economizer shall be provided in accordance with Section 503.4.1.

TABLE 506.5.1(2)
HVAC SYSTEMS MAP

CONDENSER COOLING SOURCE[a]	HEATING SYSTEM CLASSIFICATION[b]	STANDARD REFERENCE DESIGN HVC SYSTEM TYPE[c]		
		Single-zone Residential System	Single-zone Nonresidential System	All Other
Water/ground	Electric resistance	System 5	System 5	System 1
	Heat pump	System 6	System 6	System 6
	Fossil fuel	System 7	System 7	System 2
Air/none	Electric resistance	System 8	System 9	System 3
	Heat pump	System 8	System 9	System 3
	Fossil fuel	System 10	System 11	System 4

a. Select "water/ground" if the proposed design system condenser is water or evaporatively cooled; select "air/none" if the condenser is air cooled. Closed-circuit dry coolers shall be considered air cooled. Systems utilizing district cooling shall be treated as if the condenser water type were "water." If no mechanical cooling is specified or the mechanical cooling system in the proposed design does not require heat rejection, the system shall be treated as if the condenser water type were "Air." For proposed designs with ground-source or groundwater-source heat pumps, the standard reference design HVAC system shall be water-source heat pump (System 6).

b. Select the path that corresponds to the proposed design heat source: electric resistance, heat pump (including air source and water source), or fuel fired. Systems utilizing district heating (steam or hot water) and systems with no heating capability shall be treated as if the heating system type were "fossil fuel." For systems with mixed fuel heating sources, the system or systems that use the secondary heating source type (the one with the smallest total installed output capacity for the spaces served by the system) shall be modeled identically in the standard reference design and the primary heating source type shall be used to determine *standard reference design* HVAC system type.

c. Select the standard reference design HVAC system category: The system under "single-zone residential system" shall be selected if the HVAC system in the proposed design is a single-zone system and serves a residential space. The system under "single-zone nonresidential system" shall be selected if the HVAC system in the proposed design is a single-zone system and serves other than residential spaces. The system under "all other" shall be selected for all other cases.

TABLE 506.5.1(3)
SPECIFICATIONS FOR THE STANDARD REFERENCE DESIGN HVAC SYSTEM DESCRIPTIONS

SYSTEM NO.	SYSTEM TYPE	FAN CONTROL	COOLING TYPE	HEATING TYPE
1	Variable air volume with parallel fan-powered boxes[a]	VAV[d]	Chilled water[e]	Electric resistance
2	Variable air volume with reheat[b]	VAV[d]	Chilled water[e]	Hot water fossil fuel boiler[f]
3	Packaged variable air volume with parallel fan-powered boxes[a]	VAV[d]	Direct expansion[c]	Electric resistance
4	Packaged variable air volume with reheat[b]	VAV[d]	Direct expansion[c]	Hot water fossil fuel boiler[f]
5	Two-pipe fan coil	Constant volume[i]	Chilled water[e]	Electric resistance
6	Water-source heat pump	Constant volume[i]	Direct expansion[c]	Electric heat pump and boiler[g]
7	Four-pipe fan coil	Constant volume[i]	Chilled water[e]	Hot water fossil fuel boiler[f]
8	Packaged terminal heat pump	Constant volume[i]	Direct expansion[c]	Electric heat pump[h]
9	Packaged rooftop heat pump	Constant volume[i]	Direct expansion[c]	Electric heat pump[h]
10	Packaged terminal air conditioner	Constant volume[i]	Direct expansion	Hot water fossil fuel boiler[f]
11	Packaged rooftop air conditioner	Constant volume[i]	Direct expansion	Fossil fuel furnace

For SI: 1 foot = 304.8 mm, 1 cfm/ft^2 = 0.0004719, 1 Btu/h = 0.293/W, °C = [(°F) -32/1.8].

a. **VAV with parallel boxes:** Fans in parallel VAV fan-powered boxes shall be sized for 50 percent of the peak design flow rate and shall be modeled with 0.35 W/cfm fan power. Minimum volume setpoints for fan-powered boxes shall be equal to the minimum rate for the space required for ventilation consistent with Section 503.4.5, Exception 5. Supply air temperature setpoint shall be constant at the design condition.

b. **VAV with reheat:** Minimum volume setpoints for VAV reheat boxes shall be 0.4 cfm/ft^2 of floor area. Supply air temperature shall be reset based on zone demand from the design temperature difference to a 10°F temperature difference under minimum load conditions. Design airflow rates shall be sized for the reset supply air temperature, i.e., a 10°F temperature difference.

c. **Direct expansion:** The fuel type for the cooling system shall match that of the cooling system in the proposed design.

d. **VAV:** Constant volume can be modeled if the system qualifies for Exception 1, Section 503.4.5. When the proposed design system has a supply, return or relief fan motor 25 horsepower (hp) or larger, the corresponding fan in the VAV system of the standard reference design shall be modeled assuming a variable speed drive. For smaller fans, a forward-curved centrifugal fan with inlet vanes shall be modeled. If the proposed design's system has a direct digital control system at the zone level, static pressure setpoint reset based on zone requirements in accordance with Section 503.4.2 shall be modeled.

e. **Chilled water:** For systems using purchased chilled water, the chillers are not explicitly modeled and chilled water costs shall be based as determined in Sections 506.3 and 506.5.2. Otherwise, the standard reference design's chiller plant shall be modeled with chillers having the number as indicated in Table 506.5.1(4) as a function of standard reference building chiller plant load and type as indicated in Table 506.5.1(5) as a function of individual chiller load. Where chiller fuel source is mixed, the system in the standard reference design shall have chillers with the same fuel types and with capacities having the same proportional capacity as the proposed design's chillers for each fuel type. Chilled water supply temperature shall be modeled at 44°F design supply temperature and 56°F return temperature. Piping losses shall not be modeled in either building model. Chilled water supply water temperature shall be reset in accordance with Section 503.4.3.4. Pump system power for each pumping system shall be the same as the proposed design; if the proposed design has no chilled water pumps, the standard reference design pump power shall be 22 W/gpm (equal to a pump operating against a 75-foot head, 65-percent combined impeller and motor efficiency). The chilled water system shall be modeled as primary-only variable flow with flow maintained at the design rate through each chiller using a bypass. Chilled water pumps shall be modeled as riding the pump curve or with variable-speed drives when required in Section 503.4.3.4. The heat rejection device shall be an axial fan cooling tower with two-speed fans if required in Section 503.4.4. Condenser water design supply temperature shall be 85°F or 10°F approach to design wet-bulb temperature, whichever is lower, with a design temperature rise of 10°F. The tower shall be controlled to maintain a 70°F leaving water temperature where weather permits, floating up to leaving water temperature at design conditions. Pump system power for each pumping system shall be the same as the proposed design; if the proposed design has no condenser water pumps, the standard reference design pump power shall be 19 W/gpm (equal to a pump operating against a 60-foot head, 60-percent combined impeller and motor efficiency). Each chiller shall be modeled with separate condenser water and chilled water pumps interlocked to operate with the associated chiller.

f. **Fossil fuel boiler:** For systems using purchased hot water or steam, the boilers are not explicitly modeled and hot water or steam costs shall be based on actual utility rates. Otherwise, the boiler plant shall use the same fuel as the proposed design and shall be natural draft. The standard reference design boiler plant shall be modeled with a single boiler if the standard reference design plant load is 600,000 Btu/h and less and with two equally sized boilers for plant capacities exceeding 600,000 Btu/h. Boilers shall be staged as required by the load. Hot water supply temperature shall be modeled at 180°F design supply temperature and 130°F return temperature. Piping losses shall not be modeled in either building model. Hot water supply water temperature shall be reset in accordance with Section 503.4.3.4. Pump system power for each pumping system shall be the same as the proposed design; if the proposed design has no hot water pumps, the standard reference design pump power shall be 19 W/gpm (equal to a pump operating against a 60-foot head, 60-percent combined impeller and motor efficiency). The hot water system shall be modeled as primary only with continuous variable flow. Hot water pumps shall be modeled as riding the pump curve or with variable speed drives when required by Section 503.4.3.4.

g. **Electric heat pump and boiler:** Water-source heat pumps shall be connected to a common heat pump water loop controlled to maintain temperatures between 60°F and 90°F. Heat rejection from the loop shall be provided by an axial fan closed-circuit evaporative fluid cooler with two-speed fans if required in Section 503.4.2. Heat addition to the loop shall be provided by a boiler that uses the same fuel as the proposed design and shall be natural draft. If no boilers exist in the proposed design, the standard reference building boilers shall be fossil fuel. The standard reference design boiler plant shall be modeled with a single boiler if the standard reference design plant load is 600,000 Btu/h or less and with two equally sized boilers for plant capacities exceeding 600,000 Btu/h. Boilers shall be staged as required by the load. Piping losses shall not be modeled in either building model. Pump system power shall be the same as the proposed design; if the proposed design has no pumps, the standard reference design pump power shall be 22 W/gpm, which is equal to a pump operating against a 75-foot head, with a 65-percent combined impeller and motor efficiency. Loop flow shall be variable with flow shutoff at each heat pump when its compressor cycles off as required by Section 503.4.3.3. Loop pumps shall be modeled as riding the pump curve or with variable speed drives when required by Section 503.4.3.4.

h. **Electric heat pump:** Electric air-source heat pumps shall be modeled with electric auxiliary heat. The system shall be controlled with a multistage space thermostat and an outdoor air thermostat wired to energize auxiliary heat only on the last thermostat stage and when outdoor air temperature is less than 40°F.

i. **Constant volume:** Fans shall be controlled in the same manner as in the proposed design; i.e., fan operation whenever the space is occupied or fan operation cycled on calls for heating and cooling. If the fan is modeled as cycling and the fan energy is included in the energy efficiency rating of the equipment, fan energy shall not be modeled explicitly.

TABLE 506.5.1(4)
NUMBER OF CHILLERS

TOTAL CHILLER PLANT CAPACITY	NUMBER OF CHILLERS
≤ 300 tons	1
> 300 tons, < 600 tons	2, sized equally
≥ 600 tons	2 minimum, with chillers added so that no chiller is larger than 800 tons, all sized equally

For SI: 1 ton = 3517 w.

TABLE 506.5.1(5)
WATER CHILLER TYPES

INDIVIDUAL CHILLER PLANT CAPACITY	ELECTRIC CHILLER TYPE	FOSSIL FUEL CHILLER TYPE
≤ 100 tons	Reciprocating	Single-effect absorption, direct fired
> 100 tons, < 300 tons	Screw	Double-effect absorption, direct fired
≥ 300 tons	Centrifugal	Double-effect absorption, direct fired

For SI: 1 ton = 3517 w.

506.6 Calculation software tools. Calculation procedures used to comply with this section shall be software tools capable of calculating the annual energy consumption of all building elements that differ between the *standard reference design* and the *proposed design* and shall include the following capabilities.

1. Computer generation of the *standard reference design* using only the input for the *proposed design*. The calculation procedure shall not allow the user to directly modify the building component characteristics of the *standard reference design*.

2. Building operation for a full calendar year (8760 hours).

3. Climate data for a full calendar year (8760 hours) and shall reflect *approved* coincident hourly data for temperature, solar radiation, humidity and wind speed for the building location.

4. Ten or more thermal zones.

5. Thermal mass effects.

6. Hourly variations in occupancy, illumination, receptacle loads, thermostat settings, mechanical ventilation, HVAC equipment availability, service hot water usage and any process loads.

7. Part-load performance curves for mechanical equipment.

8. Capacity and efficiency correction curves for mechanical heating and cooling equipment.

9. Printed *code official* inspection checklist listing each of the *proposed design* component characteristics from Table 506.5.1(1) determined by the analysis to provide compliance, along with their respective performance ratings (e.g., *R*-value, *U*-factor, SHGC, HSPF, AFUE, SEER, EF, etc.).

506.6.1 Specific approval. Performance analysis tools meeting the applicable subsections of Section 506 and tested according to ASHRAE Standard 140 shall be permitted to be *approved*. Tools are permitted to be *approved* based on meeting a specified threshold for a jurisdiction. The *code official* shall be permitted to approve tools for a specified application or limited scope.

506.6.2 Input values. When calculations require input values not specified by Sections 502, 503, 504 and 505, those input values shall be taken from an *approved* source.

CHAPTER 6

REFERENCED STANDARDS

This chapter lists the standards that are referenced in various sections of this document. The standards are listed herein by the promulgating agency of the standard, the standard identification, the effective date and title, and the section or sections of this document that reference the standard. The application of the referenced standards shall be as specified in Section 107.

AAMA

American Architectural Manufacturers Association
1827 Walden Office Square
Suite 550
Schaumburg, IL 60173-4268

Standard reference number	Title	Referenced in code section number
AAMA/WDMA/CSA 101/I.S.2/A c440—05	Specifications for Windows, Doors and Unit Skylights	402.4.4, 502.4.1

AHRI

Air Conditioning, Heating, and Refrigeration Institute
4100 North Fairfax Drive
Suite 200
Arlington, VA 22203

Standard reference number	Title	Referenced in code section number
210/240—03	Unitary Air-Conditioning and Air-Source Heat Pump Equipment	Table 503.2.3(1), Table 503.2.3(2)
310/380—93	Standard for Packaged Terminal Air-conditioners and Heat Pumps	Table 503.2.3(3)
340/360—2000	Commercial and Industrial Unitary Air-conditioning and Heat Pump Equipment	Table 503.2.3(1), Table 503.2.3(2)
365—02	Commercial and Industrial Unitary Air-conditioning Condensing Units	Table 503.2.3(6)
440—05	Room Fan-coil	503.2.8
550/590—98	Water Chilling Packages Using the Vapor Compression Cycle—with Addenda	Table 503.2.3(7)
560—00	Absorption Water Chilling and Water Heating Packages	Table 503.2.3(7)
840—1998	Unit Ventilators	503.2.8
13256-1 (2004)	Water-source Heat Pumps—Testing and Rating for Performance—Part 1: Water-to-air and Brine-to-air Heat Pumps	Table 503.2.3(2)
1160—2004	Performance Rating of Heat Pump Pool Heaters	Table 504.2

AMCA

Air Movement and Control Association International
30 West University Drive
Arlington Heights, IL 60004-1806

Standard reference number	Title	Referenced in code section number
500D—07	Laboratory Methods for Testing Dampers for Rating	502.4.5

ANSI

American National Standards Institute
25 West 43rd Street
Fourth Floor
New York, NY 10036

Standard reference number	Title	Referenced in code section number
Z21.10.3—01	Gas Water Heaters, Volume III - Storage Water Heaters with Input Ratings Above 75,000 Btu per Hour, Circulating Tank and Instantaneous—with Addenda Z21.10.3a-2003 and Z21.10.3b-2004	Table 504.2
Z21.13—04	Gas-fired Low Pressure Steam and Hot Water Boilers	Table 503.2.3(5)
Z21.47—03	Gas-fired Central Furnaces	Table 503.2.3(4)
Z83.8—02	Gas Unit Heaters and Gas-Fired Duct Furnaces—with Addendum Z83.8a-2003	Table 503.2.3(4)

ASHRAE

American Society of Heating, Refrigerating and Air-Conditioning Engineers, Inc.
1791 Tullie Circle, NE
Atlanta, GA 30329-2305

Standard reference number	Title	Referenced in code section number
119—88 (RA 2004)	Air Leakage Performance for Detached Single-family Residential Buildings	Table 405.5.2(1)
140—2007	Standard Method of Test for the Evaluation of Building Energy Analysis Computer Programs	506.6.1
146—1998	Testing and Rating Pool Heaters	Table 504.2
ANSI/ASHRAE/ACCA Standard 183—2007	Peak Cooling and Heating Load Calculations in Buildings Except Low-rise Residential Buildings	503.2.1
13256-1 (2005)	Water-source Heat Pumps—Testing and Rating for Performance—Part 1: Water-to-air and Brine-to-air Heat Pumps (ANSI/ASHRAE/IESNA 90.1-2004)	Table 503.2.3(2)
90.1—2007	Energy Standard for Buildings Except Low-rise Residential Buildings (ANSI/ASHRAE/IESNA 90.1-2007) . . . . 501.1, 501.2, 502.1.1, Table 502.2(2)	
ASHRAE—2001, 2005	ASHRAE Handbook of Fundamentals . . . . 402.1.4, Table 405.5.2(1)	
ASHRAE—2004	ASHRAE HVAC Systems and Equipment Handbook-2004	503.2.1

ASME

American Society of Mechanical Engineers
Three Park Avenue
New York, NY 10016-5990

Standard reference number	Title	Referenced in code section number
PTC 4.1 - 1964 (Reaffirmed 1991)	Steam Generating Units	Table 503.2.3(5)

ASTM

ASTM International
100 Barr Harbor Drive
West Conshohocken, PA 19428-2859

Standard reference number	Title	Referenced in code section number
C 90—06b	Specification for Load-bearing Concrete Masonry Units.	Table 502.2(1)
E 283—04	Test Method for Determining the Rate of Air Leakage Through Exterior Windows, Curtain Walls and Doors Under Specified Pressure Differences Across the Specimen 402.4.5, 502.4.2, 502.4.8	

CSA

Canadian Standards Association
5060 Spectrum Way
Mississauga, Ontario, Canada L4W 5N6

Standard reference number	Title	Referenced in code section number
101/I.S.2/A440—08	Specifications for Windows, Doors and Unit Skylights. 402.4.4, 502.4.1	

DOE

U.S. Department of Energy
c/o Superintendent of Documents
U.S. Government Printing Office
Washington, DC 20402-9325

Standard reference number	Title	Referenced in code section number
10 CFR Part 430, Subpart B, Appendix E (1998)	Uniform Test Method for Measuring the Energy Consumption of Water Heaters	Table 504.2
10 CFR Part 430, Subpart B, Appendix N (1998)	Uniform Test Method for Measuring the Energy Consumption of Furnaces and Boilers . . . . Table 503.2.3(4), Table 503.2.3(5)	

DOE—continued

| 10 CFR Part 431, Subpart E 2004 DOE/EIA—0376 | Test Procedures and Efficiency Standards for Commercial Packaged Boilers . . . . Table 503.2.3(6) |
| (Current Edition) | State Energy Prices and Expenditure Report . . . . .405.3, 506.2 |

ICC

International Code Council, Inc.
500 New Jersey Avenue, NW
6th Floor
Washington, DC 20001

Standard reference number	Title	Referenced in code section number
IBC—09	International Building Code® 201.3, 303.2, 402.2.9
IFC—09	International Fire Code® 201.3
IFGC—09	International Fuel Gas Code® 201.3
IMC—09	International Mechanical Code® 503.2.5, 503.2.6, 503.2.7.1, 503.2.7.1.1, 503.2.7.1.2, 503.2.9.1, 503.3.1, 503.4.5
IPC—09	International Plumbing Code® 201.3
IRC 09	International Residential Code® . . . . .201.3, 403.2.2, 403.6, 405.6.1, Table 405.5.2(1)

IESNA

Illuminating Engineering Society of North America
120 Wall Street, 17th Floor
New York, NY 10005-4001

Standard reference number	Title	Referenced in code section number
90.1—2007	Energy Standard for Buildings Except Low-rise Residential Buildings . . . . .501.1, 501.2, 502.1.1, Table 502.2(2)

NFRC

National Fenestration Rating Council, Inc.
6305 Ivy Lane, Suite 140
Greenbelt, MD 20770

Standard reference number	Title	Referenced in code section number
100—04	Procedure for Determining Fenestration Product U-factors—Second Edition 303.1.3
200—04	Procedure for Determining Fenestration Product Solar Heat Gain Coefficients and Visible Transmittance at Normal Incidence—Second Edition 303.1.3
400—04	Procedure for Determining Fenestration Product Air Leakage—Second Edition. 402.4.2, 502.4.1

SMACNA

Sheet Metal and Air Conditioning Contractors National Association, Inc.
4021 Lafayette Center Drive
Chantilly, VA 20151-1209

Standard reference number	Title	Referenced in code section number
SMACNA—85	HVAC Air Duct Leakage Test Manual. 503.2.7.1.3

UL

Underwriters Laboratories Inc.
333 Pfingsten Road
Northbrook, IL 60062-2096

Standard reference number	Title	Referenced in code section number
727—06	Oil-fired Central Furnaces . . . . .Table 503.2.3(4)
731—95	Oil-fired Unit Heaters—with Revisions through February 2006 . . . . .Table 503.2.3(4)

US—FTC

United States - Federal Trade Commission
600 Pennsylvania Avenue NW
Washington, DC 20580

Standard reference number	Title	Referenced in code section number
CFR Title 16	R-value Rule	303.1.4

WDMA

Window and Door Manufacturers Association
1400 East Touhy Avenue, Suite 470
Des Plaines, IL 60018

Standard reference number	Title	Referenced in code section number
AAMA/WDMA/CSA 101/I.S.2/A440—08	Specifications for Windows, Doors and Unit Skylights	402.4.4, 502.4.1

INDEX

EDITORIAL CHANGES – SECOND PRINTING

Page 28, Table 402.1.3: Note d has been deleted.

Page 39, Table 502.2(1): column 3, row 8, now reads . . . R-5.7ci^c

Page 39, Table 502.2(1): column 4, row 8, now reads . . . R-5.7ci^c

Page 39, Table 502.2(1): column 8, row 8, now reads . . . R-9.5ci

Page 39, Table 502.2(1): column 8, row 10, now reads . . . R-13 + R-7.5

Page 39, Table 502.2(1): column 11, row 11, now reads . . . R-13 + R-3.8

Page 39, Table 502.2(1): column 12, row 11, now reads . . . R-13 + R-7.5

Page 39, Table 502.2(1): column 12, row 13, now reads . . . R-7.5c

Page 41, Table 502.3: column 2, row 7, now reads . . . 1.20

Page 73, IESNA: Standard reference number now reads . . . 90.1—2007

EDITORIAL CHANGES – THIRD PRINTING

Page 25, Table 301.3(1): row 2 Warm-humid Definition is moved below row 4 Moist (A) Definition.

Page 25, Table 301.3(1): row 2 now reads . . . Marine (C) Definition—Locations meeting all four criteria:

1. Mean temperature of coldest month between -3°C (27°F) and 18°C (65°F)

2. Warmest month mean < 22°C (72°F)

3. At least four months with mean temperatures over 10°C (50°F)

4. Dry season in summer. The month with the heaviest precipitation in the cold season has at least three times as much precipitation as the month with the least precipitation in the rest of the year. The cold season is October through March in the Northern Hemisphere and April through September in the Southern Hemisphere.

Page 28, Table 402.1.3: Note c now reads . . . c. Basement wall U-factor of 0.360 in warm-humid locations as defined by Figure 301.1 and Table 301.1.

Page 28, Section 402.2.5: exception line 2 now reads . . . insulation requirements in Table 402.2.5 shall be permit-

Page 30, Section 402.4.2.1: line 5 now reads . . . of 50 pascals (1 psf). Testing shall occur after rough in

Page 30, Section 402.5: line 4 now reads . . . 405 shall be 0.48 in Zones 4 and 5 and 0.40 in Zones 6 through

Page 32, Section 403.9.1: line 4 now reads . . . heaters fired by natural gas or LPG shall not have continu-

Page 35, Table 405.5.2(1)—continued: column 1, line 9 now reads . . . Service water heating[h, k]

Page 42, Deletion arrow added below Section 502.4.8

Page 62, Section 506.5.2: line 3 now reads . . . mal blocks as required in Section 506.5.2.1, 506.5.2.2 or

Page 64, Table 506.5.1(1): column 2, row 12, line 2 now reads . . . Table 505.6.2(2). Areas and dimensions of tradable and

Page 76, Index P: line 5 now reads . . . Parallel Path Calculation 402.2.5

Page 77, Index V: line 2 is deleted.

EDITORIAL CHANGES – FOURTH PRINTING

Page 32, Section 404.1: line 1 now reads . . . 404.1 Lighting equipment. A minimum of 50 percent of the

Page 35, TABLE 405.5.2(1)—continued: column 2, row 2 now reads . . . Specific leakage area (SLA)^e = 0.00036 assuming no energy recovery

Page 35, TABLE 405.5.2(1)—continued: column 3, row 2, line 7 now reads . . . rate^f but not less than 0.35 ACH

Page 35, TABLE 405.5.2(1)—continued: column 3, row 5, line 4 now reads . . . element^g but not integral to the

Page 35, TABLE 405.5.2(1)—continued: column 1, row 7, now reads . . . Heating systems[h]

Page 35, TABLE 405.5.2(1)—continued: column 1, row 8, now reads . . . Cooling systems[h, j]

Page 35, TABLE 405.5.2(1)—continued: column 1, row 9, line 2 now reads . . . heating[h, k]

Page 38, TABLE 502.1.2: column 2, row 8 now reads . . . U-0.58

Page 38, TABLE 502.1.2: column 10, row 8 now reads . . . U-0.090

Page 38, TABLE 502.1.2: column 11, row 8 now reads . . . U-0.080

Page 38, TABLE 502.1.2: column 8, row 19 now reads . . . F-0.860

Page 39, TABLE 502.2(1): column 11, row 11 now reads . . . R-3.8ci

Page 39, TABLE 502.2(1): column 12, row 11 now reads . . . R-7.5ci

Page 39, TABLE 502.2(1): column 13, row 11 now reads . . . R-7.5ci

Page 41,Section 502.4.4 Hot gas bypass limitations: is renumbered, relocated and now reads . . . 503.4.7 Hot gas bypass limitations

Page 55, Section 503.4.7 Hot gas bypass limitations is added

Page 75, Index M, MOISTURE CONTROL is deleted.

EDITORIAL CHANGES – FIFTH PRINTING

Page v, Chapter 4 Residential Energy Efficiency, line 3 now reads . . . is unique for this code. In this code, a *residential building* is an R-2, R-3 or R-4 building three stories or less in height. All other R-1

Page 35, TABLE 405.5.2(1)—continued: column 1, row 9 now reads now reads . . . Service water heating[h, i]

Page 37, Section 502.2.5: line 10 now reads . . . area if the material weight is not more than 120 pounds per

Page 55, Section 503.4.7: line 6 now reads . . . 503.4.7.

Page 75, Index: new entry added now reads . . . **HOT BYPASS**. 503.4.7

EDITORIAL CHANGES – SIXTH PRINTING

Page 35, TABLE 405.5.2(1)—continued: column 1, row 9 now reads . . . Service H_2O heating[h, k, i]

EDITORIAL CHANGES – SEVENTH PRINTING

Page 27, TABLE 402.1.1: footnote j now reads . . . For impact rated fenestration complying with Section R301.2.1.2 of the *International Residential Code* or Section 1609.1.2 of the *International Building Code*, the maximum *U*-factor shall be 0.75 in Zone 2 and 0.65 in Zone 3.

Page 39, TABLE 502.2(1): column 8, row 10, line 2 now reads . . . R-7.5ci

Page 42, Section 503.2.3: lines 4 and 5 now reads . . . 503.2.3(4), 503.2.3(5), 503.2.3(6), 503.2.3(7) and 503.2.3(8) when tested and rated in accordance with the

Page 47, TABLE 503.2.3(5): row 1, column 4 now reads . . . MINIMUM EFFICIENCY[c, d, e]

Page 49, TABLE 503.2.3(8) is added and reads as shown.

Page 51, Section 503.2.7.1.3: Equation 5-2 now reads . . . $CL = F / P^{0.65}$

Page 51, TABLE 503.2.8: footnote b equation now reads . . .$T = r[(1+t/r)^{K/k}-1]$

Page 55, Section 503.4.5: line 3 now reads . . . 503.4.5.4 shall apply to complex mechanical systems serv-

EDITORIAL CHANGES – NINTH PRINTING

Page 56, Section 505: title now reads . . . **SECTION 505 ELECTRICAL POWER AND LIGHTING SYSTEMS (Prescriptive)**

Page 56, Section 505.1: title now reads . . . **505.1 General (Prescriptive).**

Page 56, Section 505.2: title now reads . . . **505.2 Lighting controls (Prescriptive).**

Page 59, Section 505.3: title now reads . . . **505.3 Tandem wiring (Prescriptive).**

Page 59, Section 505.4: title now reads . . . **505.4 Exit signs (Prescriptive).**

Page 60, Section 505.6: title now reads . . . **505.6 Exterior lighting. (Prescriptive).**

Page 61, Section 505.7: title now reads . . . **505.7 Electrical energy consumption. (Prescriptive).**

EDITORIAL CHANGES – TENTH PRINTING

Page 41, Section 502.4.1: line 7 now reads . . . exceed the values in Section 402.4.4.

    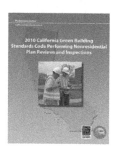

IECC TOOLS

ICC Code Resources help you learn, interpret and apply the IECC effectively

IN THE FIELD

FITS IN YOUR POCKET!
ENERGY INSPECTOR'S GUIDE: BASED ON THE 2009 INTERNATIONAL ENERGY CONSERVATION CODE® AND ASHRAE/IESNA 90.1-2007
Your ideal resource for effective, accurate, consistent, and complete commercial and residential energy provisions. This handy pocket guide is organized in a manner consistent with the inspection sequence and process for easy use on site. Increase inspection effectiveness by focusing on the most common issues relevant to energy conservation. (76 pages)
SOFT COVER #7808S09
PDF DOWNLOAD #8886P09

ENERGY EFFICIENCY CERTIFICATE STICKERS
The energy provisions in IRC® Section N1101.8 and IECC Section 401.3 require a type of certificate be installed. This easy-to-use sticker clearly lists the general insulation, window performance, and equipment efficiency details. Sold in packets of 25.
#0726S

CODE SOURCE: ENERGY CONSERVATION CODE
This value-packed resource will serve as a helpful in-the-field reference guide and as a critical component of the code enforcement and inspection process. Designed to assist field inspectors and plans examiners in the completion and performance of their duties, it will instill a solid knowledge of the practical application of the 2009 IECC and the standards set forth by the American Recovery and Reinvestment Act (ARRA).

Features:
- Lists the most common code items warranting examination for compliance.
- Organizes them in a manner that is both efficient and effective.
- Comprehensive coverage prepares building industry professionals and students for safe, accurate, and code-compliant work.
- The "quick tab" format allows easy access to critical information.
- Durable laminated pages withstand a variety of field conditions. (75 pages)
#4866S09

LEARN MORE ABOUT IECC TOOLS TODAY! 1-800-786-4452 | www.iccsafe.org/store

COMMENTARY

CODE AND COMMENTARY TOGETHER!
2009 IECC: CODE AND COMMENTARY

This convenient and informative resource contains the full text of the IECC, including tables and figures, followed by corresponding commentary at the end of each section in a single document.

- Read expert Commentary after each code section.
- Learn to apply the codes effectively.
- Understand the intent of the 2009 IECC with help from the code publisher.

The CD and download versions contain the complete Code and Commentary text in PDF. Search text, figures and tables; or copy and paste small excerpts into correspondence or reports. (245 pages)

SOFT COVER #3810S09
PDF DOWNLOAD #878P09
CD-ROM (PDF) #3810CD09

STUDY TOOLS

GREAT EXAM PREP TOOL!
2009 INTERNATIONAL ENERGY CONSERVATION CODE STUDY COMPANION

The ideal way to master the code for everyday application, to prepare for exams, or lead a training program. This Study Companion provides a comprehensive overview of the energy conservation provisions of the 2009 IECC, including the requirements for both residential and commercial energy efficiency. (188 pages)

- 10 study sessions contain key points for review, applicable code text and commentary.
- 20-question quiz at end of each study session.
- Helpful answer key lists the code reference for each question.

Great resource for Certification exams: Commercial Energy Inspector, Commercial Energy Plans Examiner, Residential Energy Inspector/Plans Examiner, or Green Building—Residential Examiner

SOFT COVER #4807S09
PDF DOWNLOAD #8787P09

FLASH CARDS: 2009 IECC

Provides code users, students and exam candidates with an effective, time-tested, easy-to-use method for study and information retention. Prepared and reviewed by code experts to ensure accuracy and quality. (60 cards)

#1821S09
BUY THE IECC STUDY COMPANION AND FLASH CARDS TOGETHER AND SAVE!
#4807BN09

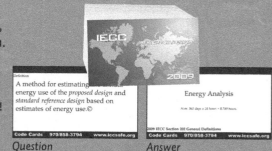

Question   Answer

LEARN MORE ABOUT IECC TOOLS TODAY! 1-800-786-4452 | www.iccsafe.org/store

 **ICC EVALUATION SERVICE**

HOW DO YOU KNOW WHEN A PRODUCT IS COMPLIANT
with codes, standards or rating systems

Tap into free resources from ICC Evaluation Service® (ICC-ES®)

- Online directory of code-compliant building products and systems

- CEU webinars, presentations and videos

- Unmatched expertise and experience in technical evaluations

ICC-ES is the most widely accepted and trusted brand and industry leader in performing technical and environmental evaluations of building products and systems.

Learn more:

www.icc-es.org
1.800.423.6587 (x42237)
es@icc-es.org

Look for the marks of conformity code officials trust:

12-06663